REPORT

T0302540

Assessing the Effectiveness of the International Counterproliferation Program

Jennifer D. P. Moroney, Aidan Kirby Winn,
Jeffrey Engstrom, Joe Hogler, Thomas Durrell-Young,
Michelle Spencer

Prepared for the Defense Threat Reduction Agency

Approved for public release; distribution unlimited

RAND NATIONAL DEFENSE RESEARCH INSTITUTE

The research described in this report was prepared for the Defense Threat Reduction Agency. The research was conducted within the RAND National Defense Research Institute, a federally funded research and development center sponsored by the Office of the Secretary of Defense, the Joint Staff, the Unified Combatant Commands, the Navy, the Marine Corps, the defense agencies, and the defense Intelligence Community under Contract W74V8H-06-C-0002.

Library of Congress Cataloging-in-Publication Data

Assessing the effectiveness of the international counterproliferation program / Jennifer D. P. Moroney ... [et al.].
p. cm.
Includes bibliographical references.
ISBN 978-0-8330-5259-9 (pbk. : alk. paper)
1. Nuclear nonproliferation—International cooperation. I. Moroney, Jennifer D. P., 1973-

JZ5675.A88 2011
327.1'747—dc23

2011034076

Published 2011 by the RAND Corporation
1776 Main Street, P.O. Box 2138, Santa Monica, CA 90407-2138
1200 South Hayes Street, Arlington, VA 22202-5050
4570 Fifth Avenue, Suite 600, Pittsburgh, PA 15213-2665
RAND URL: http://www.rand.org/
To order RAND documents or to obtain additional information, contact
Distribution Services: Telephone: (310) 451-7002;
Fax: (310) 451-6915; Email: order@rand.org

Preface

The threat posed by the proliferation of weapons of mass destruction (WMD) is among the central strategic national security challenges that the United States is facing. The breadth and scope of this challenge requires interagency cooperation, as well as coordination with a broad array of international partners. To effectively confront the threat of WMD proliferation, the United States relies on the will and capacity of its allies and international partners for assistance. This cooperation often requires investments in security cooperation programs aimed at enhancing partner capacity.

Assessing the impact of security cooperation efforts is inherently difficult. However, such assessments generate data on which more-informed decisions about program funding at all levels of government can be based. In addition to serving the needs of decisionmakers, assessments provide information to those directly involved in the planning and implementation of security cooperation programs. This report demonstrates how RAND's assessment framework, developed in previous RAND research, can be applied to combating WMD programs by illustrating its utility for the International Counterproliferation (ICP) Program.

This study draws on previous RAND research sponsored by the Defense Threat Reduction Agency (DTRA) Advanced Systems and Concepts Office, Office of the Deputy Assistant Secretary of Defense for Partnership Strategy and Stability Operations, and the U.S. Air Force and documented in the following monographs:

- Moroney, Jennifer D. P., Jefferson P. Marquis, Cathryn Quantic Thurston, and Gregory F. Treverton, *A Framework to Assess Programs for Building Partnerships*, Santa Monica, Calif.: RAND Corporation, MG-863-OSD, 2009
- Moroney, Jennifer D. P., and Joe Hogler, with Benjamin Bahney, Kim Cragin, David R. Howell, Charlotte Lynch, and S. Rebecca Zimmerman, *Building Partner Capacity to Combat Weapons of Mass Destruction*, Santa Monica, Calif.: RAND Corporation, MG-783-DTRA, 2009
- Moroney, Jennifer D. P., Joe Hogler, Jefferson P. Marquis, Christopher Paul, John E. Peters, and Beth Grill, *Developing an Assessment Framework for U.S. Air Force Building Partnerships Programs*, Santa Monica, Calif.: RAND Corporation, MG-868-AF, 2010.

RAND National Defense Research Institute

This research was sponsored by DTRA and conducted within the International Security and Defense Policy Center of the RAND National Defense Research Institute, a federally funded research and development center sponsored by OSD, the Joint Staff, the Unified Combatant Commands, the Navy, the Marine Corps, the defense agencies, and the defense Intelligence Community.

For more information on the RAND International Security and Defense Policy Center, see http://www.rand.org/nsrd/ndri/centers/isdp.html or contact the director (contact information is provided on the web page).

Contents

Figures

Tables

Summary

The ICP Program is one of many programs implemented to confront the threat of WMD proliferation. Its mission is to counter the threat of the proliferation of WMD-related materials and technologies across the borders and through the independent states of the former Soviet Union, the Baltic region, and Eastern Europe. One of the unique characteristics of the ICP Program is its interagency foundation. The U.S. Department of Defense (DoD) implements the ICP Program, in coordination with the Federal Bureau of Investigation (FBI), the U.S. Department of Homeland Security (DHS) U.S. Customs and Border Protection, and other federal agencies. The program strives to enhance the ability of host-country agencies to prevent and counter the proliferation of WMD and related items across international borders. It also aims to aid partner nations in their own efforts to deter, detect, investigate, and respond to crimes regarding WMD and related items, whether they are chemical, nuclear, radiological, or biological agents or "dual-use" items, (items that can be used for both licit and illicit purposes). The program works toward the establishment of a professional cadre of border and law enforcement and related personnel among the participating nations while delivering technical assistance via low- and high-tech equipment, training, and guidance. The ICP Program's ultimate goal is the establishment of an enduring and mutually beneficial relationship between host-country agencies and those of the U.S. government that helps to realize their shared goals of counterproliferation.[1]

Study Approach

The study team was tasked with expanding on the RAND security cooperation assessment framework prepared for DTRA in 2009 to more fully develop an analytically sound and repeatable process for program assessment, as well as demonstrating how the assessment framework can be applied to the ICP Program. First, RAND's security cooperation assessment framework was tailored to allow for an assessment of the ICP Program.[2] To ensure that we had the correct objectives, indicators, and questions for the assessment, we consulted ICP Program plans and after-action reports from past events. The team developed assessment indicators tied directly to ICP Program objectives. Second, we conducted focused discussions with key DTRA stakeholders to gain insights into the perceived value of the program from a U.S. perspective. Next, we conducted interviews with partner country officials to ensure that we obtained balanced

[1] For an overview of the history of the ICP Program, please see Appendix B.

[2] See Moroney and Hogler, 2009, pp. 69–82.

data, informed by all stakeholders. Finally, we assessed the results, determined key findings across the three case studies (Kosovo, Romania, and Georgia, discussed below), and recommended ways to improve the effectiveness and efficiency of the ICP Program, based on our review of the reports, focused discussions, and case study work.

The Assessment Framework

RAND's assessment framework includes five key elements:

- strategic guidance
- programs
- stakeholders
- authorities (including directives and instructions)
- levels of assessment that are linked with a discussion of assessment indicators (inputs, outputs, and outcomes).

Strategic Guidance

The relevance of strategic guidance, in the context of program assessment, is that it establishes the importance of developing capabilities and cooperating with partner militaries to meet U.S. strategic goals, as well as directing program managers to carry out assessments.

Programs

Within the context of the framework elements, *programs* refers to the categorization of the various types of security cooperation programs, as found in strategic guidance. This report focuses on the application of the assessment framework to a single, specific program, the ICP Program.

Stakeholders

Stakeholders refers to assessment stakeholders—persons or organizations that make decisions for, or about, a program. There are different types of stakeholders, which can be further classified depending on their particular relationship to the program.

Authorities

Authorities refers to the authorities for conducting security cooperation programs derived from the U.S. Code, DoD directives, and instructions. Title 10 of the U.S. Code is the basic authority for DoD activities; it also serves as the primary authority for many DoD security cooperation programs. Title 22 of the U.S. Code provides the basic authority for foreign assistance, including security assistance.

Five Levels of Assessment

Assessments of security cooperation programs can provide various types of valuable information regarding different aspects of the program:

- level 1: need for the program
- level 2: design and theory
- level 3: process and implementation

- level 4: outcome or impact
- level 5: cost-effectiveness (relative to other, similar programs).[3]

These five levels, which are integral to the assessment framework, can be thought of as a hierarchy, as depicted in Figure S.1.

Field Research

The study team conducted field research on three case studies—Kosovo, Romania, and Georgia—through a series of interviews with representatives from partner country agencies who had some familiarity and past exposure to the ICP Program. We supplemented these conversations with interviews with U.S. stakeholders in order to obtain a broad and balanced collection of perspectives. Appendix D of this report lists the specific interview questions. We developed these questions specifically to provide insight into the various levels of assessment: need for the program, design and theory, process and implementation, and outcomes and impact. The responses were submitted to a subjective analytic process in which the study team assigned values to the content of the response according to a three-level Likert scale: positive, negative, or somewhat positive.

Key Observations

The key observations that emerged from the three partner country cases collectively represent a combination of both best practices and challenges for the ICP Program. We organize these observations according to the five assessment levels.

Figure S.1
The Five Levels of Assessment

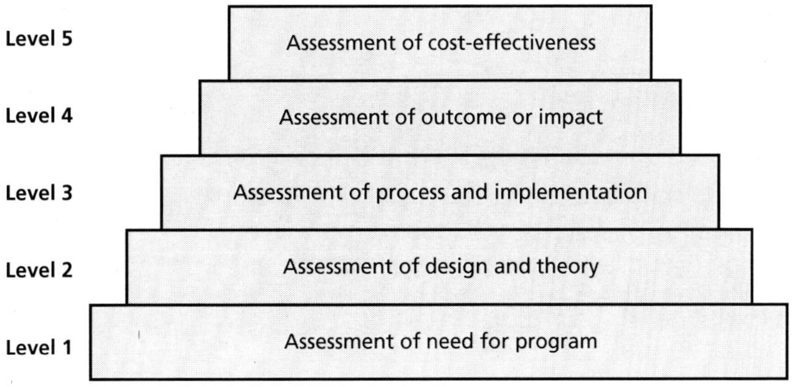

RAND *TR981-S.1*

[3] We did not include a cost-effectiveness assessment in this study, as the cost-effectiveness assessment is a measure of relative benefit based on cost and requires comparison with similar programs. We did not have access to the required budgetary information to carry this out.

Need for Program

- The ICP Program could benefit from a more rigorous methodology for partner selection to ensure that partners receiving assistance are the most relevant from a strategic perspective.
- The ICP Program's length and scope is not transparent for partner countries, leaving them unsure about the commitment of ICP Program officials for the future.

Design and Theory

- The limited partner involvement in threat assessments could serve as the basis for informed dialogue about the entire program.
- The phase construct is not analytically sound and conveys misleading information about the sophistication of partners.
- Mechanisms for partner-feedback collection are limited to course participants; input from senior-level officials is not captured effectively, which could limit the impact of the program.
- High turnover in action officers undermines the ICP Program's effectiveness at a management level and hinders relationship-building efforts.

Process and Implementation

- ICP Program courses are largely supply driven, which is not a good indicator of the need for the program.
- The curriculum has become more dynamic over time, indicating the program's ability to adapt to changing circumstances.
- Logistical support is strong and should continue to be outsourced in-country.
- Interagency coordination on the U.S. side is improving but requires additional effort to develop a whole-of-government approach to combating WMD proliferation.
- Prebriefs and after-action reviews at the event level lack assessment-relevant content, leaving follow-on efforts without real context for where assistance should be focused in a particular country.

Outcome and Impact

- ICP Program effects are difficult to determine due to the abundance of related activities, making the program difficult to assess in certain countries.
- The ICP Program fosters interagency cooperation within partner nations; often, ICP Program events are the only time when these agencies train and exercise together.
- Partners have developed informal train-the-trainer mechanisms, making the program an ideal model for other, related assistance programs.

Recommendations

This section presents several suggestions of how the program could be strengthened. To help clarify their utility and enable their effective application, the recommendations are organized according to four major segments of the program life cycle: planning, execution, sustainment, and assessment.

Planning

Adjustments to planning will enhance ICP Program effectiveness.

- Solicit feedback from U.S. and key partner country stakeholders during the ICP Program planning process.
- Expand engagement with senior partner country officials after events; keeping them informed will only help to ensure the longer-term impact of ICP Program events.
- Enhance trip prebriefs with ICP Program–relevant data so that the team understands the strategic and operational context for working with that partner country.

Execution

Execution could be enhanced through oversight mechanisms and in-country advocates.

- Identify and continually engage U.S. ICP Program advocates in-country; they are the best source of on-the-ground information about the program's impact.
- Consider forming a high-level oversight body to oversee the ICP Program.
- Consider ways to mitigate lack of relevant expertise among action officers, their rapid turnover, and ways to ease the transition with the partner nation from one action officer to another.

Sustainment

The ICP Program's impact can be sustained through creative approaches.

- Develop plans for sustaining the program's impact over the long term, particularly with more-mature partner countries.
- Use the new ICP Program portal as an engagement tool; this tool will be a useful way to stay engaged between ICP Program events.
- Introduce a recurring senior-level working group with partner countries to keep the senior officials firmly engaged.

Assessment

Strengthen assessment through focus on measurable objectives and indicators.

- Include objectives, indicators, and results in after-action reports to set a benchmark.
- Identify assessment roles for stakeholders to ensure that data are being collected and assessed in a more systematic way.

Acknowledgments

The authors are grateful for the support of many individuals. In particular, the support of our sponsors, LTC Bryan Lee and Ruth Keipp, was extremely valuable during the course of the entire study effort, as were the contributions of the various ICP Program action officers who eagerly shared their insights and experiences with us. We also wish to thank the embassy teams in Georgia, Romania, and Kosovo who helped to facilitate our field research. We appreciate the insights and perspectives shared with us by long-time ICP Program oversight officials in the Office of the Under Secretary of Defense for Policy, as well as those provided by the ICP Program support team at Computer Sciences Corporation.

The authors are extremely grateful for the thoughtful and thorough reviews of Damon Wilson of the Atlantic Council and Michael Shurkin of RAND. Finally, we wish to thank our communications analyst, Michael Neumann; and Melissa McNulty for administrative support. Any errors are solely the responsibilities of the authors.

Abbreviations

AAR	after-action report
ANCEX	National Agency for Export Controls
AOR	area of responsibility
BCCO	Combating Organized Crime Brigade
BWPP	Biological Weapons Proliferation Prevention
COCOM	combatant command
CTR	Cooperative Threat Reduction
DHS	U.S. Department of Homeland Security
DoC	U.S. Department of Commerce
DoD	U.S. Department of Defense
DOE	U.S. Department of Energy
DTRA	Defense Threat Reduction Agency
EXBS	Export Control and Related Border Security Assistance
FBI	Federal Bureau of Investigation
FCP	functional campaign plan
FSU	former Soviet Union
FY	fiscal year
GBSLE	Georgia Border Security and Law Enforcement
GEF	Guidance for Employment of the Force
HEU	highly enriched uranium
ICE	Immigration and Customs Enforcement
ICP	International Counterproliferation
NATO	North Atlantic Treaty Organization

OSD	Office of the Secretary of Defense
OUSD(P)	Office of the Under Secretary of Defense for Policy
PPP	Proliferation Prevention Program
SA/LW	Small Arms/Light Weapons
SLD	Second Line of Defense
SMS	Special Materials Service
TCP	theater campaign plan
WMD	weapons of mass destruction

CHAPTER ONE

Introduction

In recent years, the threat posed by the proliferation of weapons of mass destruction (WMD) has been recognized as one of the central strategic national security challenges that the United States is facing.[1] The breadth and scope of this challenge demands significant interagency cooperation, as well as extensive coordination with a broad array of international partners. The illicit transfer of WMD components, technologies, specialized equipment, or dual-use technologies by state, substate, or nonstate actors represents one of the chief concerns associated with the WMD threat. It is impossible for the United States to police the entire globe in the hopes of eliminating these trafficking threats. As a result, the United States must rely on the will and capacity of its allies and international partners to assist in thwarting the efforts of those who seek to traffic such materials and substances. Building the capacity of foreign partners around the world specifically to combat the proliferation and transit of WMD and related materials is not a new endeavor for the U.S. government. Traditionally focused on the former Soviet Union (FSU), U.S. assistance programs in this area tend to focus on

- training and equipping foreign militaries
- training and equipping foreign civilian agencies
- securing WMD facilities and improving infrastructure.

Some of these efforts have expanded to other regions that carry a risk of WMD transshipment, including Southeastern Europe, the Middle East, and Latin America. For example, the Cooperative Threat Reduction (CTR) program is assisting Albania's efforts to eliminate its chemical weapons stockpiles and is extending its reach to other regions of the world, including the Middle East. Moreover, since 1998, the Defense Threat Reduction Agency (DTRA) International Counterproliferation (ICP) Program has provided law enforcement and border security training and equipment to 22 countries. Other efforts are aimed at strengthening legal authorities, securing WMD materials, and enhancing accountability.

The U.S. military is equipped with capabilities readily transferable to partners dealing with WMD threats. Three such programs designed to transfer capabilities are the CTR Chemical Weapons Destruction program, the CTR Proliferation Prevention Program (PPP), and the ICP Program, which, together, provide the resources to train and equip foreign militaries and civilians to eliminate WMD, secure WMD facilities, and enhance border security. However, partner capacity is not built exclusively by the military. Civilian agencies can and do engage

[1] See Commission on the Prevention of Weapons of Mass Destruction Proliferation and Terrorism, 2008, and Weapons of Mass Destruction Commission, 2006.

nonmilitary counterparts, such as border guards, customs officials, and other frontline security services, in security cooperation. Some examples of these programs include the Export Control and Related Border Security Assistance (EXBS) program, the U.S. Department of Energy (DOE) Second Line of Defense (SLD) program, and the U.S. Department of Homeland Security (DHS) Container Security Initiative.

Despite a plethora of U.S. government programs and initiatives, there has not been an adequate assessment of their overall effectiveness in improving a partner country's capacity. Assessing the impact of security cooperation efforts is inherently difficult yet extremely important to provide stakeholders at all levels of government with effective tools to determine what aspects of these investments are productive and in which areas refinement and revision are required.

Those who plan and execute security cooperation activities intuitively know whether they have gained ground with respective partner nations as a result of their individual efforts. At the most basic level, officials often assert that the relationship with the partner country is simply "better" than it was prior to the execution of the activity. Although these assertions might be true, it is difficult to empirically validate such a general, subjective sense of accomplishment, especially to senior U.S. Department of Defense (DoD) policymakers and to Congress. At present, to the extent that assessments are done, they are conducted largely by the organization that executed the program. These assessments are thus subject, no matter how carefully carried out, to bias and lack of objectivity on the part of the assessors. Moreover, self-assessment is even less convincing when it is conducted by program managers, planners, and executers who often rotate rapidly, in as little as a year, for some military officers. Due to this transience, they might not have the long-term experience necessary to fully understand and evaluate the effectiveness of a program in a specific country.

DoD decisions regarding the allocation of resources rely, nonetheless, on assessments to establish programmatic priorities. Objective assessments provide important data points on which meaningful discussions about program funding can be based. Assessments provide the basis for decisions to continue, expand, or cut back existing programs. Without insight into the extent to which programs are achieving their goals, informed decisions regarding how resources should be allocated are not possible.

In addition to serving the needs of high-level decisionmakers, assessments provide critical information to those directly involved in the planning and implementation of security cooperation programs. Ultimately, quality assessment of these programs contributes to improved decisionmaking at all stages, including oversight, planning, management, resourcing, and execution, increasing both the effectiveness and the efficiency of such efforts. This report provides a framework for thinking about, planning for, and implementing security cooperation assessments at the program and country levels, using the ICP Program, managed by DTRA, as a test case.

Challenges of Assessment

Although a broad consensus exists regarding the importance of working with partners to pursue the goals of counterproliferation, there remain many questions and challenges concerning how best to measure the effectiveness of programs dedicated to this end. First, although there is strategic guidance directing the program manager to undertake such assessments, as discussed in Chapter Two, the guidance falls short of providing specific instructions for carrying out the assessments, what form they should take, and the process that should be used to collect, analyze, and validate the data. The result is that program managers who are serious about conducting assessments must develop their own assessment methods. Another central challenge associated with assessment is the difficulty of achieving a necessary level of objectivity. In many circumstances, those most intimately familiar with programs and, as such, best able to carry out assessments are involved in the planning or implementation of such programs and thereby placed in the position of self-assessment. Yet another common challenge faced by those attempting to measure the returns generated by any given program is that progress toward the ultimate long-term goals (outcomes), which best reflect success or failure, are often more difficult to measure than the initial investments (inputs) and more-immediate results (outputs). As these inputs and outputs are more easily quantified, they serve as the dominant metrics in assessment efforts.

Despite all of these challenges, however, conducting assessments at the program level remains an essential process to guide decisionmaking. Previous RAND work on assessment has explored the importance of such assessment at great length. One recent report, *A Framework to Assess Programs for Building Partnerships*, discusses the key reasons that program-level assessments are essential: First, it is at the program level that the most-important decisions are made about continuing, expanding, or cutting programs and resources devoted to security cooperation; second, assessments at that level bring to bear the different authorities, roles, and responsibilities of multiple security cooperation stakeholders; third, programs provide insight into requirements across countries and regions; and finally, a program-level assessment provides insight into how well a given stakeholder is achieving its objectives as measured against strategic goals. These considerations are reflected in DoD guidance, which requires program-level assessments by the combatant commands (COCOMs), Services, and defense-support agencies.[2]

The International Counterproliferation Program

The ICP Program is one of several U.S. government programs confronting the threat of WMD proliferation, as discussed above. Its mission is to counter the threat of the proliferation of WMD-related materials and technologies across the borders and through the independent states of the FSU, the Baltic region, and Eastern Europe. A unique characteristic of the ICP Program is its interagency foundation. DoD implements the ICP Program in coordination with the Federal Bureau of Investigation (FBI), DHS's U.S. Customs and Border Protection, and other federal agencies. The program strives to enhance the ability of host-country agencies to prevent and counter the proliferation of WMD and related items across international borders. It also aims to aid partner nations in their own efforts to deter, detect, investigate, and respond to crimes regarding WMD and related items, whether chemical, nuclear, radiological, or biological agents or "dual-use" items (items that can be used for both legitimate industrial and weapon-program purposes). The program works toward the establishment of a professional cadre of border and law enforcement and related personnel among the participating nations while delivering technical assistance via low- and high-tech equipment, training, and

---

[2] See Moroney, Marquis, et al., 2009, pp. 9–10. See also Moroney, Hogler, et al., 2010, and Marquis et al., 2006.

guidance. The ICP Program's ultimate goal is the establishment of an enduring and mutually beneficial relationship between host-country agencies and those of the U.S. government that helps to realize their shared goals of counterproliferation.[3]

Study Objectives and Analytical Approach

RAND was asked by DTRA to assist in developing and implementing an analytically sound and repeatable process for assessment, as well as to demonstrate how the assessment framework could be applied to the ICP Program. The study included three main tasks. Task 1 focused on tailoring RAND's security cooperation assessment framework to assess the ICP Program. The study team reviewed DTRA plans and reports from past ICP Program events. We then conducted focused discussions with key DTRA stakeholders to gain insights into the perceived value of the program from a U.S. perspective. The team developed assessment indicators tied directly to ICP Program objectives.

Task 2 focused on conducting the assessment itself using the country-specific case studies of Kosovo, Romania, and Georgia as a proof of concept for the utility of the assessment framework. We observed DTRA ICP Program events in-country and conducted focused discussions with key partner country officials to ascertain the value of the program to them. Finally, task 3 considered ways to institutionalize the assessment approach for the ICP Program.

This report demonstrates how RAND's assessment framework can be applied to combating WMD programs by illustrating its utility for the ICP Program. The application of this assessment framework to the ICP Program demonstrates the utility of the assessment framework as a theoretical construct as applied to a specific program in need of assessment. The assessment itself is not exhaustive, as it was carried out in the context of several limitations. For example, the findings are based on observations drawn from only three case studies. Likely, through subsequent case studies, or even with the selection of alternative case studies, the findings and key observations identified might significantly differ. Although the assessment strives to uncover important, practical, and actionable findings to aid the future performance of the ICP Program, the broader and more enduring value of this study and this report might be found in their demonstration of both how the assessment framework can be applied to a specific program, integrating program-specific objectives, and how the assessment findings can be used to inform future action with respect to program design and management.

Organization of the Report

This report is organized as follows. Chapter Two discusses the six key elements of RAND's assessment framework and how it was adapted from RAND's previous work on assessing building-partnership programs for combating WMD. The use of the ICP Program strategic guidance to derive specific metrics and indicators is discussed, as is the development of research questions (for both partners and U.S. stakeholders) to elicit information regarding the program. Lastly, the process for analyzing the data in order to determine macrolevel observations and microlevel issues is explained.

---

[3] For an overview of the history of the ICP Program, see Appendix B.

Chapter Three illustrates how the assessment framework was tailored and applied to assess the ICP Program relative to three country case studies: Kosovo, Romania, and Georgia. The background of each case study is discussed, followed by an explanation of the methodology used to conduct field research and analyze the results. Then, key observations identified by the study team's field research are explored.

Chapter Four presents the study team's recommendations, which are organized by the four phases of the program life cycle: planning, execution, assessment, and sustainment. This chapter aims to translate the team's observations into practical and actionable suggestions that promise to strengthen the ICP Program and pave the way for new practices that will aid future assessment efforts.

CHAPTER TWO

Adapting the Assessment Framework to Assess the International Counterproliferation Program

Security cooperation program managers conduct assessments of their programs for a variety of reasons, both internal and external to the program. Department-level assessment requirements, such as those levied by the Office of the Secretary of Defense (OSD) or the Joint Staff, are external to programs and require program managers to gather and provide data in response to these outside requests. Similarly, financial and budgetary audits, inspector general inspections, and acquisition reviews can all be considered forms of external assessment requirements. Internal requirements, on the other hand, are those driven by the program manager in an attempt to understand the effectiveness of the program and make adjustments that might improve its implementation or design. As any program manager can attest, none of these assessment requirements can be fulfilled easily if a process for routinely collecting and assessing data is not already in place.

This chapter describes such a process. The study team was tasked with expanding on the RAND security cooperation assessment framework prepared for DTRA in 2009 to more fully develop an analytically sound and repeatable process for program assessment. Specifically, DTRA requested that the team demonstrate how the assessment framework could be applied to the ICP Program.[1] The chapter begins with a brief review of the elements of RAND's assessment framework described in previous work. These elements include strategic guidance, authorities, programs, and stakeholders. Next, the chapter turns to the context in which assessments are conducted: the planning and execution process. Finally, the chapter concludes with a detailed presentation of the process in which the assessment framework elements are applied to the actual conduct of assessments. To do this, the chapter introduces and illustrates the five levels of assessment.

The assessment process laid out here can support both the external and internal assessment requirements that the program manager faces. RAND's assessment framework contains five main elements: strategic guidance, programs, stakeholders, authorities (including relevant directives and instructions), and the levels of assessment, which are linked with a discussion of assessment indicators. The way in which the assessment framework was tailored specifically to the ICP Program will be explored throughout this chapter with particular focus on the fifth element of the assessment framework, the five levels of assessment: the need for the program, design and theory, process and implementation, outcomes and impact, and cost-effectiveness. The levels of assessment are linked with a discussion of assessment indicators (e.g., inputs, outputs, and outcomes).

---

[1] See Moroney and Hogler, 2009, pp. 69–82.

Elements of the Assessment Framework

Each element of RAND's assessment framework is presented briefly in this section, beginning with a discussion of strategic guidance, which serves as the source of objectives for programs and is the standard against which program effectiveness is measured. Closely related to guidance are authorities, which provide the basis for being able to conduct security cooperation activities. Next, the term *program* is defined, providing the starting point for the discussion of program assessment. Next, the various assessment roles of stakeholders are described. These four elements are closely linked to previous RAND work on security cooperation assessments but have been tailored here so as to better fit the needs of the ICP Program.[2]

Strategic Guidance

Strategic guidance regarding security cooperation derives from a series of key documents produced at the national and departmental levels[3] of government—these include the National Security Strategy,[4] the National Defense Strategy,[5] the National Military Strategy,[6] the Guidance for Employment of the Force (GEF),[7] and the COCOM theater campaign plans (TCPs) and functional campaign plans (FCPs). OSD's GEF also directly informs the development of plans by the Services and the defense-support agencies, such as DTRA. It is important that assessments be closely linked to strategic guidance. This guidance informs decisions regarding country plans and partner prioritization, which, to a considerable extent, helps to determine the focus on ICP Program resources. All of these strategic documents emphasize the importance of developing capabilities among partner militaries to help achieve U.S. strategic goals.[8] Figure 2.1 depicts the relationship of the various levels of strategic guidance as it pertains to security cooperation.

The GEF drives the development of the COCOM and Service and defense-support agency plans, which, in turn, lead to the development of specific country plans. The GEF explicitly calls for annual assessments of programs to be delivered to the Office of the Under Secretary of Defense for Policy (OUSD(P)), including in the area of security cooperation. The GEF also contains an assessment annex, which provides slightly more-detailed guidance relevant to conducting assessments, but here, again, the guidance does not specify how such assessments should be conducted.

Despite the existing guidance, comprehensive assessments are not being conducted on a consistent basis. In 2007, a U.S. Government Accountability Office report examining U.S. counterproliferation efforts found, in relation to State Department programs, that the

2 See Moroney, Marquis, et al., 2009; Moroney and Hogler, 2009; and Moroney, Hogler, et al., 2010.

3 *National* refers to documents prepared at the White House level (e.g., the National Security Strategy). *Departmental* refers to documents prepared at the cabinet level (e.g., National Defense Strategy, prepared by DoD).

4 White House, 2010.

5 DoD, 2008.

6 U.S. Joint Chiefs of Staff, 2011.

7 DoD, 2008.

8 See DoD, 2010.

> impact of U.S. bilateral assistance to strengthen countries' abilities to counter nuclear networks is uncertain because U.S. agencies do not consistently assess the results of this assistance. The impact of this assistance is difficult to determine because the Department of State did not evaluate either (1) the proliferation risk for all of the countries in which network activities are alleged to have occurred or (2) the results of its assistance efforts.[9]

Authorities

Briefly, *authorities* refers to the legislative basis for conducting security cooperation programs derived from the U.S. Code, DoD directives, and instructions. Title 10 of the U.S. Code is the basic authority for DoD activities; it also serves as the primary authority for many DoD security cooperation programs. Title 22 of the U.S. Code provides the basic authority for foreign assistance, including security assistance.[10] Within the existing authorities provided under the U.S. Code, there are often subsequent DoD directives and instructions and even program-level instructions that lay out specific roles and responsibilities for the program, down to the stakeholder and resource-management levels. For example, the DTRA ICP Program is governed by a memorandum of agreement, which was signed in 1997.[11]

Figure 2.1
Security Cooperation Guidance Flow

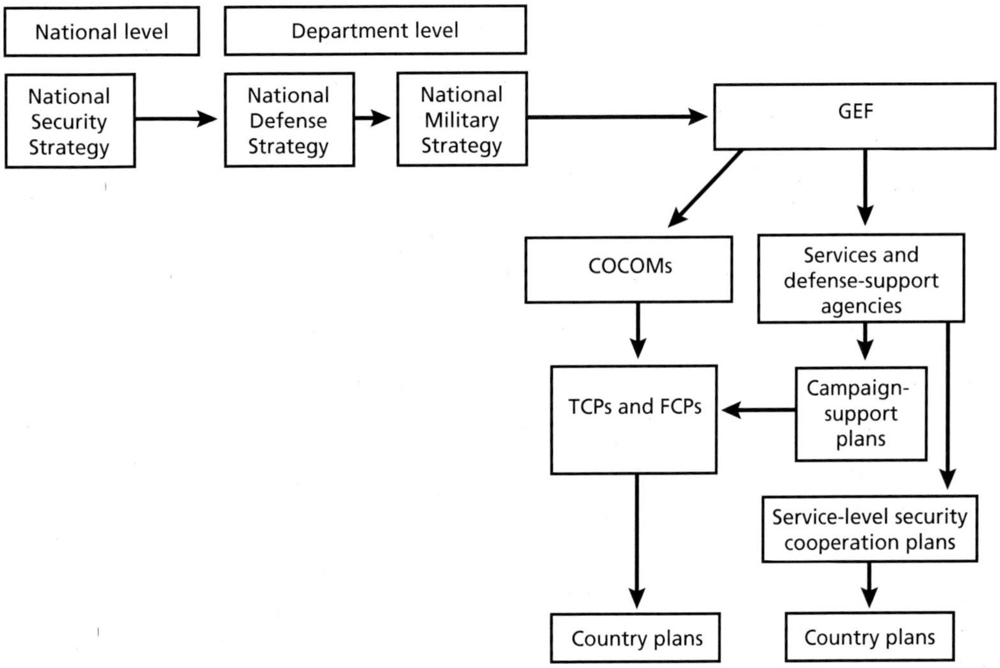

RAND *TR981-2.1*

[9] U.S. Government Accountability Office, 2007. The ICP Program was acknowledged in this report but not evaluated as part of the study. The report did comment on the fact, however, that, at that time, the ICP Program was not conducting program assessments.

[10] For a more extensive discussion of authorities, see Moroney, Marquis, et al., 2009.

[11] See Appendix B for more information on the history of the ICP Program.

Programs

Within the context of the framework elements, a program is the unit of analysis for this study. Programs have a manager, reporting requirements to a higher-level authority, and, typically, assigned resources. This analysis is focused on assessment at the program level, for several reasons. First, it is at the program level at which the most-important decisions about continuing, expanding, or cutting programs and resources devoted to security cooperation are made. Second, assessments at that level bring to bear the different authorities, roles, and responsibilities of multiple security cooperation stakeholders. Third, programs provide insight into requirements across countries and regions, not just in one specific COCOM area of responsibility. This is an important fact for defense-support agencies, such as DTRA, and also for the military Services, as they all have global (vice regional- or country-specific) responsibilities. Finally, a program-level assessment can provide insight into how well a given stakeholder is achieving its objectives against strategic goals.[12]

Stakeholders

In the context of a program, *stakeholder* refers to a person or organization that either affects or is affected by the activity conducted within that program. The concept of stakeholders is important when it comes to program assessment.[13] Security cooperation stakeholders can be thought of as the various organizations that are involved with overseeing, planning, or implementing and executing security cooperation programs. Often, a single program has many stakeholders. It is important for different stakeholders to be involved in the assessment process, as they all offer unique perspectives, and the more stakeholders that are involved, the better the chance the assessment will be objective. Assessment stakeholders are persons or organizations that make decisions for, or about, a program. For example, ancillary stakeholders, those that are curious about a program or are downstream and affected by it but do not make or contribute to decisions, are not assessment stakeholders. They might, however, provide data that contribute to assessments. In previous work, RAND researchers have identified four key roles of functional assessment that can belong to different stakeholders: data collector, reviewer, assessor, and integrator.[14]

Data Collector. According to standards set by the assessor organization, the data-collector organization is responsible for collecting and aggregating data from internal and external sources for a particular level of programmatic assessment.

Assessor. This organization is responsible for setting data-collection standards for a particular kind of programmatic assessment and for evaluating programs using methods suitable for the types of assessment it performs.

Reviewer. This organization is responsible for helping assessors develop data-collection standards and evaluation methods appropriate for the level of assessment for which they are responsible. It is also responsible for conducting periodic inspections or audits to ensure that program assessments are being properly executed.

---

12 For a full list of the security cooperation program categories, see Moroney, Marquis, et al., 2009.

13 See Moroney, Marquis, et al., 2009, p. 31.

14 See Moroney, Marquis, et al., 2009, pp. 47–48.

Integrator. This entity is responsible for organizing and synthesizing programmatic assessments to meet DoD requirements for the GEF, Capability Portfolio Management, and the Planning, Programming, Budgeting, and Execution process.

Although certain parties can play more than one role in the assessment process, it is essential that different parties carry out the assessor and reviewer functions in order to guarantee an adequate level of objectivity.

The Planning and Execution Process

In practice, assessments should be intimately tied to the planning and execution of security cooperation activities.[15] One way to think about planning and execution is that they make up an iterative process that relies on a feedback loop; security cooperation activities are executed as planned, and further plans are informed by the results of previous execution. Planning and execution of ICP Program activities are a good example of how this process works. Considering the planning portion of the process first, there are three steps that the ICP Program security cooperation planner should undertake:

1. identifying capabilities and desired end states relative to the WMD threat
2. working with potential partners
3. identifying relevant security cooperation ways and means.

Identifying Capabilities and Desired End States Relative to the WMD Threat

This initial step in the planning process is very important, as it forms the basis for how and why a program is implemented. Understanding the nature of the threat is the first element; planners identify capabilities and objectives relative to that threat. Although, in many cases, planners must carefully work through this step before moving forward with the rest of the planning process, this step is essentially done for ICP Program planners. Strategic guidance for the ICP Program has already been developed by OUSD(P) and disseminated to DTRA in a memorandum that describes specific program objectives. Moreover, this policy memorandum describes the WMD threat that is being addressed by the program and specifies the types of capabilities that the ICP Program will deliver to partner countries.

Working with Potential Partners

Once threats, objectives, and capabilities have been determined, two questions remain to be answered in the planning process regarding partner selection and the nature of activities to be undertaken:

1. With which partners should the program work ("who")?
2. What activities should the program undertake with them ("how")?

The selection of partners can, in some cases, be determined by program managers, but, to a large extent, these decisions are influenced directly by priorities identified at the level of

[15] See much more detailed discussion of the planning and execution process in Moroney and Hogler, 2009.

strategic guidance.[16] An answer to the "who" question should be based on the relevance of the potential partners, the U.S. relationship with the potential partners, and the partners' willingness and capacity to work with the United States in the context of combating WMD proliferation.[17] In the case of the ICP Program, a potential partner's relevance could be indicated, for example, by its geographical proximity to a WMD-related threat or its proximity to ungoverned or undergoverned spaces; these partners are candidates to be recipients of ICP Program assistance. Alternatively, some potential partners, such as advanced allies or regional leaders, might be selected because of their relevance in assisting the United States in meeting its ICP Program objectives in a region.

An examination of indicators of capacity and willingness can reveal much about how appropriate a potential partner might be and what types of capabilities and capacities might be created by a program like the ICP Program. For example, previous participation in U.S. nonproliferation assistance programs, such as the SLD, EXBS, or any number of other, similar programs might indicate a potential partner's general willingness to work with the United States through the ICP Program. A review of a potential partner's ability to absorb training and equipment transfers; its current capabilities, such as WMD response teams; and previous training or equipment provided by the United States or other like-minded countries can provide an indication of current capacities. Much of these data can be obtained from the embassy teams and from open-source and classified intelligence sources. This information, in turn, can help the ICP Program planner tailor training and equipment to enhance or sustain current capacities; build new, desirable capabilities; or, in some cases, decide that engagement is simply not appropriate.

Identifying Relevant Security Cooperation Ways and Means

Once the potential partners are selected, the security cooperation planner must consider the available ways (or activities) and means that will be used. This is the "how" question. For the ICP Program, this is a relatively straightforward step because it has been determined to be the ICP Program, and the ways and activities that the program uses are training, equipment transfer, and seminars. Tailoring these to a specific partner is intimately tied to the objective identified in the first step.

Executing the Program

Finally, once the planning process is complete, the ICP Program activities are executed with the partner countries. If the planning and execution process were to stop here, the ICP Program planner would be left to develop a new process every time another activity is conducted with the partner country, without having the knowledge of ICP Program effectiveness from a historical perspective. But, as mentioned earlier, the process is an iterative loop, dependent on assessments to build on successes and address failures. Moreover, such assessments will inform the planner as to how and when to improve, sustain, or terminate assistance. Best practices and challenges associated with executing the ICP Program will be dealt with in greater detail in Chapter Three. The remainder of this chapter is dedicated to examining the actual conduct of

---

[16] Ultimately, guidance regarding the appropriateness of given partners for inclusion in the ICP Program (or any other similar program) should be derived from COCOM priorities.

[17] Moroney and Hogler, 2009, p. 39.

assessment, using the five levels of assessment, in various contexts, as these levels of assessment were specifically applied to carry out the ICP Program assessment.

Five Levels of Assessment

Assessments of security cooperation programs can provide various types of valuable information regarding different aspects of the program,[18] as divided into the following five levels in our framework:

- level 1: the need for the program
- level 2: design and theory
- level 3: process and implementation
- level 4: outcome and impact
- level 5: cost-effectiveness (relative to other, similar programs).

Level 1, which assesses the need for the program or activity, focuses on the problem to be solved or goal to be met and identifies the population to be served and the kinds of services that might contribute to a solution. Once a need assessment establishes that there is a problem or policy goal worth pursuing,[19] different solutions can be considered. This is where theory connects various types of security cooperation activities to strategic goals. Assessment at the design and theory level (level 2) focuses specifically on the design of a policy or program. Level 3 assessments, process and implementation, deal with the execution of the elements prescribed by the design and theory at level 2. Assessments of outcome and impact at level 4 measure the changes (often long term) that ultimately resulted from the program's existence. Last, level 5, cost-effectiveness, assesses the relative value gained from the program as compared with that gained from other, similar efforts.

These five aspects, which are integral to the assessment framework, can be thought of as a hierarchy,[20] as depicted in Figure 2.2.

A positive assessment at a higher level relies on positive assessments at the lower levels. By *positive*, we mean that the assessment reveals that associated objectives are being met. Accordingly, problems noted at a higher level are likely linked to problems existing at lower levels. This means, for example, that *a cost-effectiveness assessment cannot be completed without the feedback from assessments at the other four levels.*[21] Moreover, if a cost-effectiveness assessment were to reveal a problem, information from the lower levels must be examined in order to fully understand the root cause of the problem. The following section links the planning and execution process described in the first section of this chapter with the five levels of assessment.

---

[18] See Moroney, Hogler, et al., 2010, pp. 55–61, for a detailed discussion of the five levels of assessment.

[19] In the case of the ICP Program, this need has been determined.

[20] Adapted from Rossi, Lipsey, and Freeman, 2004.

[21] As agreed with the sponsor, we did not perform a cost-effectiveness assessment in this study, as the cost-effectiveness assessment is a measure of relative benefit based on cost and requires comparison with similar programs. A comparative analysis of this kind was beyond the scope of the study.

Figure 2.2
The Five Levels of Assessment

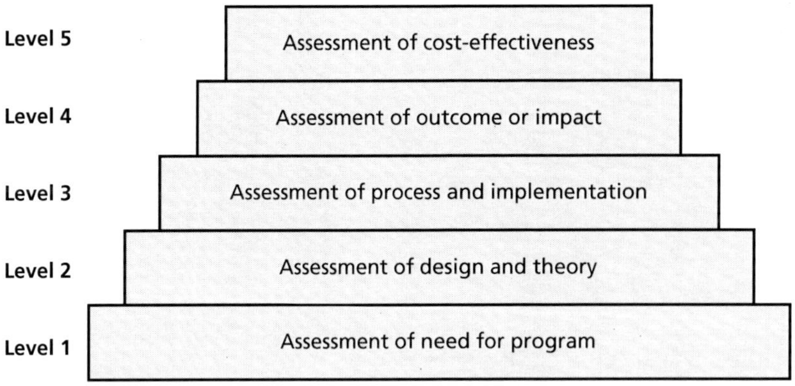

RAND *TR981-2.2*

Linking the Planning Process to the Assessment Levels

The planning and execution process of the ICP Program can be linked to the assessment levels, allowing the assessment framework to be operationalized for this specific program. Figure 2.3 illustrates the planning and execution process with the five assessment levels overlaid. This figure was developed to assist program managers in conceiving of the assessment levels as an ongoing dynamic process, directly connected to program execution.

In the top left of the figure is a box labeled "Ongoing needs assessment." Within this box are the two key planning actions that contribute to this type of assessment: identification and validation of objectives and capabilities, and consideration of WMD threats. The next box, on

Figure 2.3
Linking Assessments to Planning and Execution

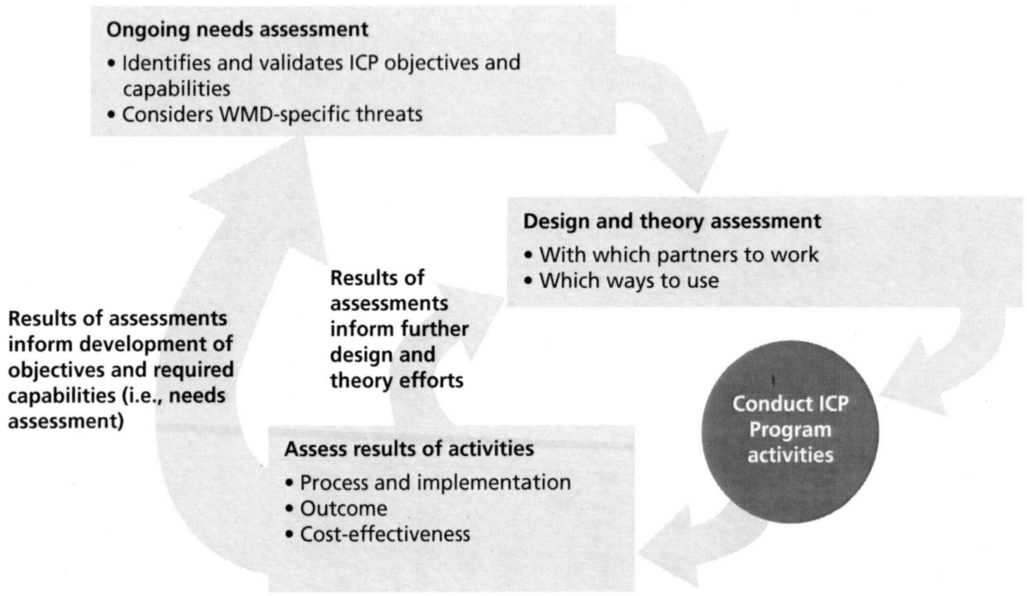

RAND *TR981-2.3*

the right side of the figure, is labeled "Design and theory assessment." Within this box are the two planning actions that feed into this type of assessment: the "who" and the "how" described in the previous section. In short, assessing at the "need for the program" and the "design and theory" levels is directly linked to the planning portion of the process. The next three assessment levels are linked to the execution portion of the process.

The bubble in the lower right of the figure is labeled "Conduct ICP Program activities." This bubble represents the execution of the ICP Program plan. The results of this plan execution must be captured and used to feed to the remaining three levels of assessment: implementation, outcomes, and cost-effectiveness. The results of these assessments, in turn, feed into the further assessment of design and theory and the need for the program, closing the loop of this iterative process.

Putting the Assessment Framework into Practice

This section describes how levels 3, 4, and 5 (process and implementation, outcome or impact, and cost-effectiveness) are supported.

Using the two documents that lay out the mission and strategy for the ICP Program (specifically, the OSD Strategic Policy Guidance[22] and the ICP Program Seven-Year Strategy[23]), the study team identified the program objectives (outcome and output), as well as the indicators. The OSD guidance document starts by describing the program's mission: "to build capabilities of participating nations to prevent proliferation of weapons of mass destruction (WMD) and related materials across their borders" (p. 1), which we interpret as the outcome objective that is being measured in level 4, outcome or impact. The OSD memo also provides goals, which we interpret as output objectives (they are lesser in scope and more easily measured). Output objectives are measured in level 3, process and implementation.

The team assumed that each course has its own output objectives (and these vary by course) and that, as with any program, there are administrative and funding output objectives for the ICP Program. Administrative and funding output objectives can be understood as management objectives. The seven output objectives appear in Figure 2.4.

An important aspect of the nested nature of the assessment levels is that each of these output objectives contributes to meeting the overall ICP Program outcome objective. In other words, if they are successfully met, then one can assume that the outcome objective has been achieved. Moreover, objectives, such as funding, also inform the fifth level, cost-effectiveness. The guidance documents also describe deliverables (tasks)—e.g., plans, funding documents—that can also be thought of as indicators of progress toward output objectives. That is, they can be linked to the output objectives. The implication for conducting assessments is that the indicators can be identified and, subsequently, analyzed to determine whether the output objectives are being achieved. Appendix C outlines the full table of ICP Program indicators and the associated output objectives they measure. The utility of the list in Appendix C is that ICP Program planners now have an initial list of indicators and output objectives. The list can

---

22 Principal Deputy Assistant to the Secretary of Defense for Global Security Affairs, undated.

23 ICP Program, 2008.

Figure 2.4
International Counterproliferation Program Output Objectives

Strategic objectives

1. Enhancing participants' ability to prevent, detect, deter, interdict, identify, and investigate the trafficking of WMD-related materials within their territory and across their borders.
2. Promoting the establishment of a professional cadre of law and border enforcement personnel dedicated to proliferation prevention, including their national security, military, and defense sectors, and encouraging effective interagency processes.
3. Emphasizing a regional approach to WMD-related border security issues, where feasible.
4. Establishing long-term relationships between U.S. government agencies and agencies of participating countries with responsibility for WMD security.

Management objectives

5. Meet administrative requirements.
6. Meet individual course objectives.
7. Manage long-term funding effectively.

RAND *TR981-2.4*

and should be modified, but the point is that it provides a useful starting point in a longer-term plan to institutionalize an ICP Program assessment process.

Conclusion

Program assessment is a difficult but a very important process to inform decisionmaking at the policy and program levels. The framework presented here is intended to aid in the assessment process by breaking down a complex task into its various components. The discussion of the framework illustrates how to take a theoretical structure and tailor it to a specific program by linking it with concrete program objectives. In the next chapter, we examine how questions related to each of the indicators result in data that can support each of the five assessment levels. The questions, developed for both ICP Program managers and partner countries, were used consistently in each of three case studies to illustrate how the framework can be operationalized.

CHAPTER THREE

Applying the Framework to Assess the International Counterproliferation Program

The ICP Program has been operating for approximately 15 years but, as of yet, has not undergone a comprehensive assessment at the program level. After operating mainly in Eurasia, the ICP Program is now considering the regions into which it might effectively expand its engagement. Furthermore, the ICP Program is an interagency program with unique authorities in terms of training and equipment provision abroad. Gaining insight into the program's successes and challenges to date will benefit ICP Program managers, enabling them to make informed decisions regarding the program's future and the establishment of new international partnerships. All of these factors make an ICP Program assessment a timely and worthwhile exercise.

This chapter illustrates how the assessment framework outlined in Chapter Two was applied to the ICP Program for the purposes of this study. First, the chapter discusses the approach used to select the case studies, followed by a brief background on the three countries' histories with the ICP Program. Next, the approach taken in conducting field research is explained, followed by a presentation of the findings and key observations, organized according to the levels of assessment.

Approach to Case Study Selection

In close coordination with the sponsor, through several consultations, we selected three case study countries: Kosovo, Romania, and Georgia. In selecting appropriate case studies for our analysis of the ICP Program, we considered a number of different criteria. First, the team sought to identify three countries[1] whose experiences and histories with the ICP Program were sufficiently different to provide unique insights and lessons. Second, we sought to achieve some geographic diversity within the confines of Southern and Eastern Europe and Eurasia by selecting countries spanning more than one region (by selecting these three cases, the regions of the Western Balkans, the Southern Caucasus, and Southern Europe are included). Third, the team considered the importance of the availability of data and, as such, sought to include countries with extensive histories with the program (longer than five years) among the three selected. Last, the study team determined that it would be more analytically rich to study three countries with varying levels of capability achieved through the ICP Program training (i.e., the preference was to examine at least one country considered by ICP Program managers to be a high-achiever nation, as well as a nation whose capabilities currently appear more modest).

[1] Resources allocated for this study would support only three in-depth case studies.

Ultimately, Kosovo, Romania, and Georgia emerged as strong candidates. Conveniently, these three countries all had events or planning visits scheduled within the duration of the study, providing opportunities for focused discussions with partner country officials.

Case Study Histories

Each of the three case study countries is discussed briefly in this section with respect to its historical relationship with the ICP Program.

Kosovo

Kosovo is a relatively new partner within the ICP Program and, as such, provided limited historical data as a case study. We selected Kosovo despite this short history because it offers a perspective on how the ICP Program approaches partner selection and the initial steps of engagement in the current context. Over the 15-year history of the program, many practices have evolved. By selecting Kosovo, a new partner, as a case study, we were guaranteed a snapshot of current practices, to be examined in the context of the approaches taken with longer-standing ICP Program participants. Although Kosovo did not meet the criteria of having an extensive history with the program and therefore was not as data rich from the perspective of *historical data* as the two other cases, the proximity since its WMD Executive Seminar (conducted in 2010) provided the opportunity to directly access participating senior-level officials, which offered a useful complement to the other two cases, in which these officials were not available for interview.

In 2009, Kosovo had its first WMD Executive Seminar, the kick-off event that is held at the outset of a new partnership. According to Kosovo and U.S. officials, this event was a great success, enabling talk across various ministries that had rarely, if ever, occurred in the past.[2] The study team was able to interview its participants and examine how the initial steps of engagement frame what will hopefully become a positive, long-term relationship.

Although the length of Kosovo's participation in the ICP Program might be too limited to effectively evaluate the program's impact from a capacity-building standpoint, it nonetheless offers important insights into program design that the other case studies were not able to provide.

Romania

We selected Romania as a case study because the ICP Program has been engaged there for several years, but the engagement level has ebbed and flowed. Romania is generally regarded as a fairly advanced country (among ICP Program partners) and, therefore, provided sufficient historical data to be effective. Although the ICP Program's initial focus was on central Asia, it approached countries in Eastern and central Europe in 1997 to determine their level of interest in counterproliferation cooperation with the United States. Romania was one of the first partner countries to express a strong desire to partner with the United States in this area. DoD conducted policy discussions with Romanian government officials in May 1997, followed by a training needs assessment in late summer. Romanian officials were eager to participate in both customs and law enforcement–oriented training. The early focus leaned toward the ICP

[2] Discussions with Kosovo and U.S. officials, Pristina, Kosovo, December 2009.

Program's customs-related courses with hands-on training for border officials, including joint training with both Bulgaria and Moldova at their respective shared borders with Romania.

After having participated in ICP Program WMD detection and interdiction courses, and having had the benefit of resident U.S. Customs WMD advisers funded by the ICP Program, the governments of Bulgaria and Romania agreed to conduct a combined exercise at their common border ports of Ruse, Bulgaria, and Giurgiu, Romania. This was the first combined exercise the ICP Program held. The operation, named ROMBUL, was conducted from June 18 to June 21, 1998. Planning for this operation required extensive interministerial coordination in both countries, which served to raise the overall level of awareness of the counterproliferation threat. The exercise was deemed a success by the United States and the regional partners.[3]

Then, in May 1999, Bulgarian customs officials seized 10 grams of highly enriched uranium-235 (HEU) at the Ruse border crossing point. Two of the Bulgarian customs officials who made the seizure had received ICP Program training during the train-up phase immediately prior to Operation ROMBUL.[4]

Romania enjoyed extensive engagement from the ICP Program between 1997 and early 2001. During these years, more than 100 Romanian officials participated in ICP Program events. In January 2001, the ICP Program conducted a program review to assess the effects of the ICP Program training and to determine whether reforms were necessary. These reviews were conducted in only a few cases, but it is unclear whether they were deemed inefficient or were halted for other reasons. At the request of the Romanian government, additional policy consultations occurred in 2004, prior to a follow-up WMD Executive Seminar in September 2004.[5] Since 2004, most of the ICP Program efforts in Romania have focused on law enforcement training and regional events.

In addition to the joint border exercise that led to a seizure of radioactive material, the ICP Program in Romania can claim other impacts. Romania's National Agency for Export Controls (ANCEX) is the national authority for controlling the export and import of strategic goods. ANCEX leaders attended the first WMD Executive Seminar and used knowledge gained there, as well as through the Department of State's EXBS Program training, to help revise Romania's export control laws. This fact was noted at the 2004 WMD Executive Seminar by the Director of ANCEX.[6]

The ICP Program has trained approximately 450 Romanian government officials to date, many of whom have advanced within their sectors and institutionalized lessons learned through the ICP Program's training events. ICP Program engagement with Romania is in a

---

[3] Discussions with former U.S. officials working within the ICP Program at the time of initial events with Romania, March 2010.

[4] The customs port director at Ruse had attended ICP Program training, as had the director of the Bulgarian central customs chemical laboratory who analyzed the material.

[5] The policy consultations were conducted by OSD and the DTRA contractor, who met with the U.S. Country Team and with representatives from all relevant agencies in the Romanian government. These consultations gave the program the flexibility to focus its efforts where they were deemed most effective by the host country.

[6] In Romania, the police are responsible for internal security and combating organized crime and terrorism. The police force has an elite enforcement unit called the Combating Organized Crime Brigade (BCCO). The Special Materials Service (SMS) Nuclear Unit is a first-response unit within the BCCO. The SMS Nuclear Unit is the largest recipient of ICP Program equipment and training in the Romanian government.

sustainment phase.[7] The government of Romania is supportive of a strong U.S. presence in the region. This fact, evolving threats, and the long-term relationship established with Romania should all be considered important factors when determining next steps for ICP Program activities in that country.

Georgia

Georgia was selected as a relevant case study because, although it has also been involved in the ICP Program for several years, it has not reached the same level of capability as other long-standing partners. Georgia was chosen because the study team believed that it would yield sufficient data to shed light on the question of why some countries appear to excel more quickly than others, even given similar investments.

Georgia was one of the first countries outside Central Asia to be engaged by the ICP Program. The first assessment took place in late 1996, and training began in 1998. Georgia was one of the only ICP Program partner countries that requested and received two WMD Executive Seminars in consecutive years (1998 and 1999). These events helped generate the buy-in and commitment of more than 50 Georgian government officials to form the basis of long-term inter- and intragovernment cooperation.

According to former officials with whom we spoke, one of the more interesting efforts executed in Georgia involved a cooperative project between the ICP Program and the U.S. Department of Commerce (DoC). According to U.S. officials who planned and participated in the event, it was clear that such states as Georgia had little experience in formulating a legal basis for preventing and deterring proliferation.[8] Thus, the ICP Program and DoC conducted a legal forum that brought together U.S. and Georgia legal officials to evaluate existing laws, decrees, and normative acts in order to determine whether adequate legal authority existed for a comprehensive counterproliferation program. The workshop provided an opportunity for the drafting of new laws and regulations that were eventually enacted by the Georgian parliament.

As efforts to confront global terrorism escalated following the terrorist attacks on the United States in September 2001, Georgia played an integral role for the United States in undermining terrorist transit and training opportunities in the Pankisi Gorge area bordering Chechnya. ICP Program officials questioned Georgia's ability to absorb additional training and, thus, around 2002, dramatically reduced its efforts but maintained contact with Georgian officials on an ad hoc basis. The ICP Program reengaged with Georgia in late 2004 and again found a willing partner.[9] At this point, the focus of the program's efforts broadened from a largely export control and WMD-containment focus to a focus on terrorism threats and WMD transit through the Caucasus region.

The ICP Program in Georgia is one of many U.S. efforts pursuing related goals and objectives. The Georgia Border Security and Law Enforcement (GBSLE) Assistance Program began

---

[7] The sustainment phase is the final phase of the ICP Program that applies to countries that have enjoyed significant engagement (approaching or exceeding seven years) and, theoretically, whose capability is judged to have reached such a level that the ICP Program objectives now focus on sustaining the impacts achieved rather than further increasing capability.

[8] Discussions with former U.S. officials working within the ICP Program at the time of initial events with Georgia, March 2010.

[9] Discussions with former U.S. officials working within the ICP Program at the time of initial events with Georgia, March 2010.

in 1998, motivated by concerns about Georgia's ability to control its borders and ports after withdrawal of Russian border guard forces. The U.S. Department of State provides funding and policy oversight for this program, while the U.S. Customs Service (now in DHS) implements it as the executive agent. The State Department and DHS coordinate these efforts with other U.S. government agencies, including DoD, DoC, DOE, the FBI, and others attempting to ensure that these assistance efforts are effective and complementary, although, due to the plethora of border security assistance programs directed at Eurasian partners, this is not always the case.[10]

In addition, the CTR Biological Weapons Proliferation Prevention (BWPP) program has stepped up efforts to remove and destroy dual-use equipment from the Biokombinat facility outside Tbilisi. The BWPP program's Threat Agent Detection and Response program is also providing the Georgian government with several high-security diagnostic labs to detect diseases and bioagents.

DTRA has also held more than 100 arms control, border security, counterproliferation, and consequence management training and exercise activities in the Black Sea region through the Small Arms/Light Weapons (SA/LW) program and the Defense Treaty Inspection Readiness Program. The ICP Program has trained well over 600 Georgian officials through more than 30 events held in Georgia since 1998. It is important to recognize the varied programs that have operated in Georgia over the past several years, as the variety underscores the complex context in which any one program's effectiveness must be assessed.

According to press reports, there have been several attempts to smuggle HEU across Georgia's borders, the most recent of those reported occurring in April 2010.[11] Although it has not been determined whether any of the ICP Program–trained officials were directly involved in the interdictions, the intense investigations that followed each event reflected Georgia's political will and commitment to countering proliferation of WMD. ICP Program officials view Georgia's extensive participation, at both the national and regional levels, as having helped to inculcate that political will and commitment. That sentiment is shared by Georgian officials the team interviewed.[12]

Field Research

The study team reviewed DoD guidance documents and detailed after-action reports prior to conducting interviews. The study team carried out field research for each of the three case studies through a series of structured conversations with representatives from partner country agencies who had some familiarity and past exposure to ICP Program activities. It is important to note that the study team made a concerted effort to speak with partner country officials with several years of direct experience working with the ICP Program officials. Most of the partner country officials interviewed are in senior management roles. However, the team was also able to speak with several people who had received ICP Program training.

---

10 Discussions with U.S. officials in Washington, D.C., and Tbilisi, Georgia, January–February 2010.

11 Corso, 2010.

12 Discussions with ICP Program officials.

Interviews with partner country officials are supplemented by interviews with U.S. stakeholders, including U.S. embassy officials, DTRA, State Department, DOE, DHS (e.g., Coast Guard, Immigration and Customs Enforcement), and the Department of Justice (i.e., FBI). The study team also interviewed headquarters elements of these agencies, as well as OSD, prior to traveling to the partner countries. The study team targeted officials from these agencies because they also conduct related security cooperation programs that often supplement and occasionally overlap with the ICP Program, in order to obtain a broad and balanced collection of perspectives.

We structured conversations with both partner country and U.S. officials around a series of questions (see Appendix D) developed within the context of the assessment framework. We adapted the questions slightly for each audience but covered essentially the same issues. Answers to these questions provide insight into the various levels of assessment: need for a program, design and theory, process and implementation, and outcomes and impact. Partners were not able to supply meaningful information regarding the cost-effectiveness of the program, nor did we access detailed records of program costs, so, although cost-effectiveness questions were developed for illustrative purposes with respect to the framework, this level was not addressed in our field research, nor is it included in our assessment findings. Table D.1 in Appendix D summarizes the responses we received from partner country interviewees. The study team used a subjective analytic process to assign values to the content of the response according to a three-level Likert scale: positive, negative, or somewhat positive.

We limited the scale to three rankings to help reduce ambiguity. For example, if a particular administrative requirement is not being accomplished, or activities are not achieving a stated course objective, the assessor would rank those areas "negative." Similarly, if the administrative requirement or objective was clearly achieved, then the assessor would rank those areas "positive." The use of a "somewhat positive" rank, as for each of the rankings, does require expert judgment, but it was most commonly assigned in instances either in which a mixture of positive or negative responses was received from different respondents for a single question or in which a single response clearly fell somewhere in between the values of positive and negative.[13] Each case study dealt with the implementation of the ICP Program in the context of a unique political and cultural setting. Therefore, it is not surprising that responses varied across the cases for each question asked. Furthermore, meaningful data regarding each question could not be obtained in every case. The information gathered was subject to the personal experiences and insights of those interviewed. As such, the availability of information varied significantly across the three cases. However, after aggregating all the partner responses, the team identified some observations that revealed themselves in more than one of the case studies. Some observations are positive, while some identify areas for improvement. These observations are summarized in the next section and organized according to the assessment levels.

It is important to note that, to the extent possible, the study team considered the motivations of the partner country officials interviewed. Often, there is a danger that countries that are net recipients of U.S. assistance will report only the positive outcomes and will "play up"

[13] For instance, question D7, as shown in Table D.1 in Appendix D, asks whether ICP Program officials solicit data on whether their courses are designed and delivered in a fashion that is appropriate to the partner country's conditions and requirements. One of the planning visits conducted in the course of field research exhibited this type of solicitation for information, but discussions suggested that this practice had not been widely applied to all training courses and exercises. Therefore, the value assigned was "somewhat positive."

the need for additional assistance. The study team discussed possible partner country official motivations with U.S. embassy officials prior to and after those interviews.

Key Observations

The key observations that emerged from the three partner country cases collectively represent a combination of both best practices and challenges for the ICP Program. We have organized these observations according to the assessment levels to further demonstrate the value of the assessment framework as a mechanism for organizing and interpreting program performance findings. The study team's analysis focuses only on the first four levels of assessment and *excludes* an assessment of cost-effectiveness. The rationale for excluding cost-effectiveness is that, in order to assess at this level, an assessment of all related combating-WMD-proliferation security cooperation activities in a given country must also be assessed. In other words, in the case of assessing the ICP Program in Georgia, the team should also assess the need, design and theory, process and implementation, and outcome and impact of EXBS, GBSLE, SLD, CTR, and other programs in Georgia to determine overall "bang for the buck" of the ICP Program. This task was beyond the scope of our research, as it deals with cost-effectiveness.

Need for Program

The ICP Program could benefit from a more rigorous methodology for partner selection. The current process for selecting participant countries is influenced by shifting political interest and other factors, which tend to render the process ad hoc. The absence of a coherent methodology for partner selection has significant implications for the program, in that it can impede defining the nature of the relationship with a partner nation and, in turn, identifying and articulating program objectives with regard to that particular nation. Now that ICP Program managers are exploring global opportunities for new partnerships, it will be increasingly important to clearly define the process by which partner countries are selected.

RAND's previous work dealing with criteria for partner selection can serve as a helpful resource in this process. In *Building Partner Capacity to Combat Weapons of Mass Destruction*,[14] RAND researchers developed an approach for planners to characterize potential partner states according to two key measures: willingness and capacity. *Willingness* refers to the potential partner's interest in working specifically with the United States to address WMD threats, whereas *capacity* refers to a partner's various traits, including its ability to secure its own borders and facilities, provide law enforcement interdiction and investigation capacity, work with its neighbors to address threats collaboratively, and so on. Subject-matter expertise is required to understand the perspective of the partner country regarding WMD in order to accurately determine how the WMD threat is perceived, its capacity to counter the threat, and any gaps it has that could be filled appropriately by U.S. security cooperation assistance. The criteria applied to the selection of a partner should be based on assessments of the threat while also considering that country's ability to absorb the assistance, as well as its willingness to work with the United States on shared objectives.

The program's length and scope are not transparent for partner countries. Partner country officials do not currently have adequate insight into the process by which the ICP Pro-

[14] Moroney and Hogler, 2009.

gram determines the length and the scope of its engagement with them. When this issue was discussed during field research, it became apparent that partner countries had little visibility into DTRA's decisions regarding how long it would continue to provide training and exercises and what criteria might be used to conclude their involvement in the ICP Program.[15] This proves problematic for a number of reasons. First, it undermines the likelihood of obtaining honest feedback on the program from the partner country, due to concern that critical feedback could jeopardize its continued participation. It also removes any reference to timeliness that could assist the partner country in its own self-assessments of capacity increases. The official program documents refer to a seven-year time frame as the rough estimate for the duration of a partner country's involvement in the ICP Program, but the factors driving this time frame are unclear. Furthermore, the participation of some countries, including Georgia, have already exceeded the seven-year time horizon, and the ICP Program still shows no sign of diminishing its engagement.

Design and Theory

There is limited partner involvement in threat assessments. There is an opportunity for the ICP Program partner countries to become more heavily involved in the development of threat assessments. The field research conducted indicates that most partner countries currently offer very little, if any, input into the development of threat assessments. This issue can be deemed sensitive, and there appears to be a misunderstanding among both U.S. and foreign officials that to increase coordination would require the sharing of classified information. In reality, however, increased collaboration in developing threat assessments need not require disclosure of sensitive intelligence sources or means of collection.

The importance of partner involvement in the development of threat assessments is derived not so much from the need to acquire highly detailed intelligence information from the partner nation, as it is from the value gained from incorporating partner viewpoints, even if based on open-source materials. Moreover, the collaborative development of a threat assessment would be useful in that it could serve as the basis of an informed dialogue on which the entire program could be founded. Indeed, consideration might be given to developing a combined (U.S. and partner) threat assessment during the executive seminar. This process is important to establishing a shared understanding of existing threats and securing political investment and "buy-in" to the ICP Program. By not in any substantive way involving the partner nation in this foundational feature of the program, complacency and skepticism can arise regarding the true nature of the WMD threat as it pertains to any particular partner.

Ideally, partner involvement would include high-level engagement from all security-relevant ministries. In some partner nations, there might be questions regarding the level of indigenous capacity to undertake such assessments. As such, the nature and extent of partner involvement will likely vary from case to case. In the case of Georgia, we found an emerging capability to undertake threat assessments being led by the country's relatively new National Security Council, with input and involvement from across its ministries.[16] This indigenous capability is one that could be harnessed by the ICP Program to generate effective input regarding Georgia's threat environment and thus secure important political commitment.

---

[15] Discussions with government officials in Romania and Georgia.

[16] Discussions with Georgian officials, Tbilisi, Georgia, January 2010.

The phase construct requires modification. Following our discussion with ICP Program managers, we determined that DTRA's phase framework is not the most effective means for determining a country's progress through the ICP Program.[17] The current phases are as follows:

- Phase I, program initiation, includes all aspects of initial engagement.
- Phase II, program consolidation, includes the initiation of basic operational and strategic training to operational managers not included in the executive seminar and is based on a defined capacity-building road map. During this stage, any equipment transfers must be preceded by a WMD agreement with that nation.
- Phase III, regional expansion, includes delivery of equipment and intermediate WMD training, as well as the expansion of program activities (if realistic) from the target nation to multiple nations within the immediate region.
- Phase IV, program integration, includes advanced WMD training, national-level strategy engagement, and cooperative regional activities with multilateral organizations, such as the North Atlantic Treaty Organization (NATO), the Organization for Security and Co-Operation in Europe, and advanced tabletop exercises geared to achieve specific objectives.
- Phase V, program sustainment, includes minimal support from DTRA. The partner nation is expected to have attained the necessary training and equipment at this point.

In terms of actual time invested, the approximate expectation is that countries will progress through the phases as follows:

- phase I: year 1
- phase II: years 2 and 3
- phase III: years 4 and 5
- phase IV: years 5 and 6
- phase V: years 7 and beyond.

In an effort to roughly determine where countries stand with respect to these phases, courses have been loosely tied to phases, but these distinctions and classifications are not strictly applied.[18] One issue with the phase construct is that it does not help ICP Program managers assess progress among participants or even classify the countries by their relative status, as it does not encompass effective metrics. Rather, these phases are input driven: Phases can be assigned according only to the length of time the ICP Program has been engaged in a given country and the courses that have been taught there. As a result, phases are largely ineffec-

---

[17] Observation of an ICP Program working-group meeting on phases, DTRA, September 22, 2009.

[18] For example, phase I courses have been identified as Policy Visit, WMD Executive Seminar, and Training Needs Assessment. Phase II courses are Terrorism and the Proliferation of WMD, Counterproliferation Awareness, Basic Investigative Analysis, International WMD Dual Use Investigations, WMD Border Management (TBD), and Evidence Awareness. Phase III courses are Combating WMD Proliferation at Maritime Ports, Green [unprotected] Border Tracking Training, Radiological Detection and Response, Responding to WMD at Borders TTX [tabletop exercise], Crisis Incident Management, and Financial Terrorism (TBD). Phase IV courses are Crime Scene Operations, Advanced Investigative Analysis (TBD), Open Source Information and Analysis (TBD), Cyber Crime Investigations, WMD National Crisis Command Center TTX, Building a National WMD Counterproliferation Strategy (TBD), and Regional Engagement Events. Countries cannot receive courses that require the use of equipment until a WMD agreement is signed, which occurs in phase III. Phase V courses are program-sustainment activities, such as Train-the-Trainer, Alumni Association/Distance Learning, and Regional Engagement Events to be arranged.

tive as a methodological tool for grouping countries according to their capabilities. The phase construct risks conveying misleading information about the sophistication of partner countries' indigenous capabilities. If countries are permitted to progress to higher phases without demonstrating increasing capability and capacity, the utility of such a framework is severely undermined.

Part of the challenge facing the ICP Program's attempt to develop a meaningful model to measure countries' progress stems from the absence of widely accepted standards governing counterproliferation capacity. This void of criteria and standards is not unique to the ICP Program; rather, it is a challenge facing a broader set of programs throughout DTRA, as well as DoD and broader U.S. government programs focused on building capabilities to confront WMD proliferation.

ICP Program managers are likely to continue to be under pressure from senior DTRA managers and OSD policymakers to depict in an at-a-glace fashion where the countries currently engaged in the ICP Program stand according to their capabilities, and which levels of resource investment these varying levels of capabilities reflect. For this reason, ICP Program managers will likely need to explore alternative, objective-based methods of reflecting this information.

Mechanisms for partner feedback collection are limited to course participants; high-level input is not captured effectively. Feedback from partner countries provides one important metric by which to determine specific and overall effectiveness of the ICP Program. Such feedback can be solicited and received in a number of ways, both formal and informal. Currently, formal and regular feedback is derived mainly from student surveys in which information is captured on the last day of training, before course-completion certificates are handed out. Course participants are given a 31-question survey that tracks each participant's background (e.g., agency represented, years of experience, role); views of the value and quality of the course and course instruction (e.g., content, relevance, usefulness, materials, length, instruction); and thoughts about logistical aspects (e.g., meals, accommodations, audio/visual, interpretation); as well as general comments.[19] The student feedback surveys employ quantitative means of tracking student views through rating statements on a 1-to-5 scale, as well as the qualitative means of soliciting comments to add perspective and gain insight on the outcome and effectiveness of the course. This survey provides a way of determining course effectiveness and tailoring future courses to better address local issues that directly concern the partner officials. Informal feedback comes through a variety of channels. Some is relayed through country-team members, who are generally familiar with both the program's core objectives and the unique context of the partner country. Also, in several instances, senior officials in partner countries have themselves previously attended ICP Program courses and are consulted in the planning of ICP Program events. This provides another mechanism by which partner feedback can be obtained.

High turnover in action officers undermines program's effectiveness. The action officers assigned to the ICP Program often lack key expertise, which makes it difficult for them to quickly get acquainted with the program and take responsibility for a given region, including the politically sensitive task of liaising with high-level partner country officials. It is crucial for successful action officers to have interagency and foreign-affairs experience (including time

---

[19] The full collection of available student surveys from courses conducted in Romania and Georgia were provided by DTRA and reviewed by the study team.

spent living abroad and some foreign-language exposure), as well as regional and country-specific subject-matter expertise in counterproliferation policy issues, and some specialized education in the highly technical areas in which the ICP Program delivers. Often, however, new action officers do not bring this experience. Moreover, the turnover among action officers occurs every one to three years, making the lack of relevant expertise a significant program liability.[20]

The fact that most action officers are detailed from their respective military departments, as opposed to being civilian hires, presents numerous challenges. These action officers have a steep learning curve and little opportunity to benefit from their predecessors' experience. New action officers tend to lack regional, cultural, or language expertise, and, accordingly, a large section of the ICP Program's most critical asset, its personnel, is ill equipped for the task.

Process and Implementation

The ICP Program courses are largely supply driven. Because of the challenges, both resource driven and otherwise, in tailoring the ICP Program curriculum to each partner nation, all participating countries tend to receive a fairly standardized set of courses. This course lineup is determined largely by the ICP Program action officers. When questioned about their level of involvement in selecting the courses they had received, the partner nations consulted for this study expressed that they were not involved in the process, nor were they given visibility into how it functioned.[21]

As a result, course delivery tends to be more supply driven than demand driven. The fixed costs required to create a course and its associated materials are significant; therefore, courses appear to be developed in a general fashion. As many interviewees pointed out, courses often rely on a standard set of case studies in their content, but only a certain number of cases are relevant to the instruction of many of these substantive and technical areas. It is not as if most countries participating in the ICP Program have a relevant case from their own national history to tailor into course materials, but it should be recognized that, in addition to fixed costs, courses also involve variable costs and, perhaps even more importantly, opportunity costs. So, although it is both practical and expected that a certain level of standardization be in place when delivering ICP Program course content (particularly for courses that are highly technical and address such subjects as forensic methods), there are certain circumstances in which the unique needs of partner nations are better met through increased variety and customization.

The scheduling of courses could also be improved, as well as decisions regarding appropriate intervals for course repetition. These decisions should be more closely tied to the input collected from partner countries and based on their specific needs, growth, or changes in, for example, mission in their bureaucracies or high turnover in certain agencies. Currently, the scheduling of these courses is heavily influenced by the program's concerns and schedule and deconfliction across the agendas for various participating countries.

Although participating countries are relatively satisfied with the courses being offered, this does not necessarily indicate that all the courses being offered effectively match their

[20] This finding is based on discussion with DTRA senior managers and discussions with interagency partners, as well as firsthand observation of action officers' work in-country.

[21] These conclusions were based on interviews with officials in Romania and Georgia, as well as extensive interviews with ICP Program action officers who undertake the planning and course-selection process for countries in their areas of responsibility (AORs).

needs. Because many of the ICP Program's participants are relatively new to issues associated with WMD proliferation and counterproliferation, it is likely that, in many cases, they are not fully aware of what their needs are and which elements of the existing ICP Program curriculum are most relevant to their particular requirements and conditions. Furthermore, as is the case for most security cooperation programs, it is difficult to effectively solicit critical feedback from participating partners, as some partners erroneously fear that critical feedback could adversely affect their status within the program or lead to the discontinuation of assistance.

The curriculum has become more dynamic. ICP Program course content is reviewed on an annual basis and submitted to a course-review process.[22] This review process evaluates the ICP Program course curricula as a whole and also examines individual course offerings.[23] Issues that have arisen during the prior year are discussed, and changes, if deemed necessary, are made, whether through new and more-relevant examples; updated tactics, techniques, and procedures; revised structure; or increased accessibility. Courses that are judged to be duplicative are also removed at this stage. In addition, classroom instruction methods employ a number of teaching methods and techniques to actively engage students and ensure that they leave the course well informed and able to apply the knowledge they have gained. This includes regular use of interactive modules, in which ICP Program instructors perform a demonstration or hands-on exercises are conducted. Furthermore, in-class instruction uses TurningPoint® presentation software that allows instructors to assess students' level of knowledge absorption at midcourse and make adjustments where necessary.

Logistical support is strong. Valuable logistical support is provided through federal contractors that have in-country offices staffed by partner country nationals. The services provided are essential to the ICP Program's overall ability to effectively and efficiently host training, planning visits, executive seminars, and regional events in partner countries, as they arrange critical logistics, including (but not limited to) participant lodging, meals, travel, and interpretation services. Our research revealed that the in-country contractor staff knows how to efficiently provide these needs, has a deep understanding of local political structures, and often has familiarity with key senior partner country officials on a professional basis.[24]

Interagency coordination on the U.S. side is improving but requires additional effort. ICP Program managers are required to coordinate the planning and execution of its programs both with U.S. country teams within the envisaged recipient country and with the U.S. COCOM in whose geographic AOR the recipient country is located.

The RAND study team found that the ICP Program's performance is highly dependent on the receptivity of the relevant staff directorates of the applicable U.S. COCOM and whether its projects are accurately and effectively represented in that command's TCP. Moreover, as the

---

[22] ICP, 2008, Attachment 4: ICP Program Sustainment Plan, p. 64.

[23] The details of this course-review process were obtained through interviews with Computer Sciences Corporation (CSC) curriculum directors, November 30, 2009.

[24] One such essential logistical support service contracted for (and mentioned briefly earlier) is interpretation. Seeking to teach and train new methods and possibly ideas in different cultural and linguistic contexts presents some challenges. The ICP Program has minimized these challenges through effective use of translation capabilities that are currently provided through the government contractor companies, such as Technology Management Company. In each classroom, a team of at least two native-level linguists interprets presentations from English into the local language. The study team was able to observe the quality of interpretation services firsthand during the course of our research trips, in which interpretation with non–English speakers was carried out seamlessly and allowed the discussion of fairly sophisticated ideas and terminology free of obstacles.

ICP Program does not have resident staff seconded to recipient countries, the success of its projects is highly dependent on U.S. Country Team involvement (e.g., its recognition in the ambassador's Mission Strategic Plan for the country and willingness to support ICP Program activities), let alone its interest and active engagement with those ministries and organizations that have responsibility for WMD counterproliferation issues.[25] In the case of Romania, what had been a fairly robust and competent WMD-detection and response capability, due, in large part, to ICP Program efforts, has been allowed to atrophy somewhat since 2000. Compounding this situation within the Romanian government was the U.S. country team's decision not to focus heavily on this issue in Romania. Under such circumstances, the ICP Program efforts might have been undermined by a foreign government's action, or inaction, and by a U.S. country team's decision to focus on other issues with the Romanian government, perhaps to the detriment of making additional progress on combating WMD issues.

Fortunately, the study team has seen the program redoubling its efforts to improve coordination with U.S. country teams.[26] Historically, however, the program has not always oriented its attention and limited resources to best suit its communications and coordination with recipient countries. The study team has observed a pattern of ICP Program personnel largely directing their attention to the tactical-level execution of programs, often at the expense of developing and maintaining connections at the policy level in recipient countries. Again, the source of this problem is likely the personnel-management system under which the program operates. Many of its action officers are junior officers, or junior field-grade officers, many with limited experience working at the national or policy level within the U.S. government, as well as limited experience liaising with foreign governments. Thus, as can be expected, absent a clear policy requiring frequent and effective engagement at this level, action officers might not be equipped to fully exploit opportunities to ensure that their respective programs produce the best effect in a country.

The lack of high-level connections in a country can be overcome, in part, by virtue of a detailed country program plan, i.e., a road map, which the ICP Program now develops for each country at the WMD Executive Seminar.[27] The study team was unable to find evidence, however, that foreign officials are habitually briefed on these plans. The lack of a full appreciation or understanding by recipient countries of ICP Program goals could account, in part, for some countries' apparent tendency to send the wrong students to the program's courses.[28] This presents a potential waste of limited resources, as well as opportunity cost, both for the country and for the ICP Program. Means to overcome, or at least mitigate, such pitfalls must be devised. Related to ensuring that the right students are seconded to the program's courses is the fact that many of the countries in which the ICP Program works are relatively small. As such,

---

[25] Discussion with ICP Program manager, December 2009.

[26] The study witnessed extensive coordination with the Georgia Country Team during the planning visit undertaken in March 2010, in preparation for the upcoming ICP Program exercise scheduled for September 2010. Feedback from Country Team officials in separate conversations relayed that this level of engagement and coordination represented a distinct improvement over past practices.

[27] Discussion with ICP Program manager, December 2009.

[28] This is, however, only one factor contributing to this practice. Interviews with U.S. stakeholders has revealed that, often, cultural customs of rewarding employees with attendance at such courses is deeply ingrained in many regions and is not easily overcome, even with sustained efforts to communicate ICP Program goals, as well as the specific content and appropriate audience for each individual course.

redundancy of knowledgeable personnel is extremely limited, and the opportunity cost of sending a student to an ICP Program course is proportionately higher than for a larger country. This reality makes it even more incumbent on the program to meet its stated goals by ensuring that courses are targeted to the specific requirements and conditions of a recipient country.

Prebriefs and after-action reports (AARs) lack assessment-relevant content. Throughout the course of the study, the RAND team was able to observe that both the prebriefs and the AARs were lacking important content in key areas, making the ultimate goal of assessing the ICP Program more difficult. In the case of the trip prebriefs, the content, although informative on a general level, tended to lack key details regarding both the specific nature of the ICP Program's relationship to that particular country (e.g., how long the program had been operating there, recent courses that had taken place, key challenges unique to this partner) and clearly articulated objectives for any specific course or exercise. In the case of the AARs, a heavy focus on the logistical dimensions of how the course was executed tended to dominate the narrative, with little attention being paid to assessment-relevant observations, including whether the event met its objectives and the articulation of any informal feedback captured from participants.

Outcome and Impact

ICP Program impacts are difficult to determine due to the abundance of related activities. One challenge inherent in assessing the performance and impact of the ICP Program in any given country results from the complexity of divorcing the program's impacts from those of all the other efforts under way, aimed at seeking similar capacity-building goals. This was particularly apparent in the case of Georgia, which has multiple programs operating there, all seeking related ends. In addition to the long-standing presence of the ICP Program, Georgia has also benefited from support via EXBS, SLD, and the GBSLE program. Although these programs are each distinct and involve unique objectives and methods, for Georgian officials, who might have attended training courses or exercises associated with multiple of these programs over the course of many years, it is often difficult to distinguish one from another and, therefore, provide feedback tailored to a specific program.[29] Furthermore, other efforts to assess knowledge, including that demonstrated in practical exercises, or real "successes," might reflect enhanced capacity gained through a combination of investments by various programs. Although this phenomenon was particularly pronounced in Georgia, it was, to a lesser extent, evident in Romania as well.

An effort was made to mitigate confusion by conducting background research into the other programs, as well as speaking with representatives in country as a means of familiarizing ourselves with recent and ongoing efforts conducted by these programs. This helped somewhat with our filtering and organizing of partner feedback, allowing us to highlight that which pertained specifically to the ICP Program and that which gives insights into the broader spectrum of U.S. combating WMD programs. This issue not only became apparent in field research with partner country representatives but also was stressed by many U.S. government stakeholders, particularly those in-country, as a challenge.

The ICP Program fosters interagency cooperation within partner nations. One very positive theme that emerged from all three case studies is that the ICP Program has proven an

---

[29] Observations from discussions with Georgian officials, Tbilisi, Georgia, January 2010.

effective mechanism for encouraging greater interagency cooperation.[30] Improved interagency cooperation occurred at two levels: the personal level and the institutional level. First, partners expressed that, as a result of ICP Program courses and exercises, they made important contacts in other agencies and developed relationships that they might not have otherwise. For many, these personal relationships proved enduring and encouraged individuals to reach more frequently outside of their own agency. Second, interviewees expressed that they had learned more about other agencies' roles in the context of WMD-related crises and had gained an appreciation for the notion of an interagency culture. These events led them to appreciate the broader context in which their work plays a part and how they could cooperate across their governments to respond to various scenarios. The training teams conducting the ICP Program courses offer a positive model of interagency cooperation, which appears to resonate with the students from partner countries.

On a less positive note, several individuals with whom we spoke admitted that, although these experiences are slowly encouraging an interagency perspective to take root, it remains the case that the only opportunity for these types of interactions tends to be seminars, trainings, and exercises hosted by foreign countries.[31] This suggests that, although partners appreciate these opportunities that have made important contributions, the goal of promoting their own interagency cooperation for the purpose of aiding counterproliferation efforts has not fully taken root within their national governments. This was most evident in Kosovo and Georgia, and much less so in Romania.

Partners have developed informal train-the-trainer mechanisms. The ICP Program is currently in the midst of developing specially tailored train-the-trainer courses to help address the growing need for sustainment mechanisms to foster ongoing progress among its more mature partners. Interestingly, as a complement to these formal efforts, our field research uncovered that partner nations themselves have developed some of their own informal train-the-trainer mechanisms that help to ensure that the benefits of ICP Program instruction and materials are able to reach a broader audience within their agencies than merely those who have had the opportunity to physically attend the courses. Examples of these informal mechanisms are as simple as senior management having their materials from the courses reproduced and circulated to subordinates, as well as the incorporation of ICP Program content into regular internal training sessions.

Conclusion

Our findings indicate that the ICP Program is currently functioning effectively in some areas. However, as our case study analysis has revealed, opportunities exist to modify current practice in order to potentially enhance the program's impact. The following chapter examines some concrete recommendations that should help to enhance the ICP Program's efforts to reach its objectives, as well as provide practical mechanisms to aid in ongoing assessment efforts.

---

[30] Feedback from interviews conducted in all three case study countries supported this finding. Although engagement in Kosovo is still nascent, partners expressed very positive feedback on the role of the WMD Executive Seminar in exposing participants to interagency approaches.

[31] This sentiment was conveyed in discussions with Georgian officials in particular.

CHAPTER FOUR

Recommendations

Now that we have identified some of the ICP Program's strengths and weaknesses through a case study analysis, it is important to acknowledge that there are some obstacles beyond the program managers' ability to address in the short term. However, these factors can significantly affect the program's potential to reach its objectives.

This chapter suggests several options for strengthening the program that fall largely within DTRA's jurisdiction to address. To help clarify their utility and enable their effective application to the program, the recommendations are organized according to four major segments of the program life cycle: planning, execution, assessment, and sustainment. Recommendations in this chapter are examples of the fruits that can be drawn from applying the RAND assessment framework.

Adjustments to Planning Will Enhance International Counterproliferation Program Effectiveness

Solicit Information from Stakeholders During the Planning Process

ICP Program managers should continue to integrate feedback from U.S. and partner stakeholders into the *initial* planning process. This recommendation comes primarily from discussions with embassy officials during the field research. By factoring this input into the planning cycle, information from the interagency and in-country stakeholders can be sufficiently considered when selecting yearly courses for each country, in addition to budgetary constraints and cross-coordination with other partners' course agendas. Also, AARs and lessons-learned documents from previous courses should be systematically examined for any relevant insights.

ICP Program managers should engage routinely with appropriate directorates within the COCOMs. Analysts from COCOM intelligence, for example, can likely provide valuable "steering" guidance in deciding where attention is most necessary in a combating-WMD-proliferation context. To secure this helpful input, closer communication with COCOMs would be required. This might demand that ICP Program managers regularly update respective COCOMs on recent initiatives and encourage the COCOMs to relay feedback, as well as pertinent information on emerging threats and priorities. More generally, it is important for the ICP Program to have a plan to regularly reengage senior-level policymakers at regular intervals regarding the direction of the program. This will help instill confidence in senior officials regarding the program's impact and might lessen the pressure to generate overly simplistic "assessments" of where each partner country stands (e.g., phases) at any given time.

Expand Engagement with Senior Partner Country Officials

When planning ICP Program courses, exercises, and associated events in partner countries, it is important for managers to recognize that the views of the partner country's senior-level officials are critical to incorporate into the planning and assessment process. This idea was expressed by the partner officials who appeared to understand ICP Program activities extremely well. This implies the need for the program to continually engage at the permanent secretary and deputy-ministerial levels in recipient countries. Currently, these senior-level officials are actively engaged during the WMD Executive Seminar, but their involvement tends to wane after that event. The WMD Executive Seminars could be expanded to include the development of combined threat or risk assessments, on which training decisions can be made.

A key document generated by the WMD Executive Seminar is the road map that represents a consensus between the ICP Program and the partner country officials regarding the direction of the relationship. These road maps can serve as the basis for regular interaction between the ICP Program and senior leadership. They should be informed by a deeper level of analysis to better understand precisely how much technical assistance can realistically be absorbed and implemented by the partner country. Moreover, the ICP Program could benefit from a regular pattern of program-management reviews with high-level representation from the U.S. country team to review progress and discuss problematic issues.

Additionally, DTRA liaison officers can act as valuable resources for the ICP Program. The current utility of DTRA liaison officers varies from country to country, as some are not heavily engaged in the ICP Program and are consumed, instead, with the execution of other DTRA programs, such as CTR, which are substantially more resource intensive. Their value tends to be personality dependent as well. However, we have observed instances in which liaison officers are absolutely indispensable, and, as such, action officers should engage them to the greatest degree possible and seek to develop close and collaborative relationships.

Enhance Trip Prebriefs with International Counterproliferation Program–Relevant Information

The trip prebriefs held prior to each course are vital, as they serve to orient all team members to the environment and provide an overview of the current political context, as well as any recent developments. The study team's observations from participating in these briefs are that they could be made more comprehensive, with more-explicit attention given to the nature of the relationship with a particular nation via the ICP Program. These briefs should carefully examine such issues as these:

- How long has the ICP Program been engaged in this partner country?
- What types of courses have been provided in the past, and why were they chosen?
- What are the goals for engagement in this partner country, and what level of progress has been made toward them in recent months and years?
- What challenges have arisen in the program's relationship with the partner country?
- For how long is continued engagement envisioned with this partner country?
- Why was the current course or exercise planned at this time, and what are its key near-term objectives?

Furthermore, trip prebriefs offer an opportunity to introduce the U.S. team, including government officials and contractor support, to assessment metrics and indicators, even if at a general level, which will help to ensure that those who execute the events can also act as effective data collectors.

Execution Could Be Enhanced Through Oversight Mechanisms and In-Country Advocates

Identify and Continually Engage U.S. International Counterproliferation Program Advocates In-Country

To enhance its impact, the ICP Program should make a greater effort to identify key ICP Program advocates within all participating partner countries. They should be engaged on a regular basis to determine, for example, WMD-proliferation issues that might be on their minds, new partner country initiatives on the horizon, and any changes that have resulted from ICP Program training. Key individuals vary from country to country but will generally include representatives from the U.S. country team, including the legal attaché and the EXBS adviser, as they tend to be closely involved in assistance programs similar to the ICP Program. The need to engage embassy officials was particularly prominent in the Georgia case study. These individuals can provide helpful insights and advice regarding appropriate strategies for targeting course participants. Coordination with such in-country stakeholders concerning ICP Program courses and exercises will likely be at its strongest and most effective if not accompanied by frequent requests for logistical support. Our research has determined that logistical coordination is best outsourced to capable subcontractors (as is currently being done). This will ensure that U.S. in-country stakeholders are consulted primarily for their subject-matter expertise and knowledge of the unique political context of the partner country. As these relationships require significant maintenance, action officers would, ideally, remain in close touch with their in-country contacts continually throughout the year, even in the absence of an upcoming course, in order to stay informed on local developments.

Consider Forming a High-Level Oversight Body to Oversee the International Counterproliferation Program

ICP Program management officials should consider creating a high-level body to serve an advisory role, overseeing the program's operation and ensuring a maintained focus on the program's primary mission. This advisory board could perform several functions by providing practical and regional expertise, as well as an additional layer of oversight where deemed beneficial by DTRA leadership. This body would ideally be composed of senior interagency officials (e.g., from DoD, State, Justice, DHS, and DOE or some combination), possessing relevant experience and subject-matter expertise. The advisory body's authority could easily extend beyond the ICP Program to include other DTRA-managed security cooperation programs, such as SA/LW, and perhaps some of the CTR programs, such as the Proliferation Prevention Program (PPP). It is also possible that this body could support the assessment process. Another potential role for the oversight body in the assessment process is discussed in the next section.

Consider Ways to Mitigate Lack of Relevant Expertise Among Action Officers, as Well as Their Rapid Turnover

ICP Program managers must manage their personnel within the confines of difficult exogenous factors. For instance, when scheduling the rotation of personnel, it is not always possible to provide for an overlap period during which departing action officers would have sufficient time to effectively train their replacements. Furthermore, many new action officers initially lack key areas of expertise, including foreign-affairs training and exposure to the interagency. As a result, new action officers are forced to learn by doing on the job. Although many of these factors lie beyond the control of the ICP Program's management, investing time and thought into practical ways of mitigating these challenges would be worthwhile. The study team has observed that a structured management and oversight system has not been established to date that would enable the effective sharing of experiences and expertise among action officers.

The fact that ICP Program action officers have generally managed to execute such successful programs is likely a reflection of the dedication and hard work by individual action officers, as opposed to an institutionalized process to capture such essential information. A more institutionalized approach to the sharing of experience would benefit the ICP Program. Perhaps there are methods for outgoing action officers to capture key information derived from on-the-job experience and impart it to newcomers, even in the absence of an overlap in their positions. This information might differ from the formal guidelines that likely appear in training materials. These informal lessons and tips can be gleaned only from practical experience.

The International Counterproliferation Program's Impact Can Be Sustained Through Creative Approaches

Develop Plans for Sustaining the Program's Impact Over the Long Term

At the time of this writing, OUSD(P) and DTRA leadership, as well as ICP Program management, have not yet identified the best courses of action for sustaining the ICP Program's impact among mature partners. Close examination of the challenges the ICP Program faces to remaining effectively engaged with mature partner countries has implications for both the ICP Program and the wider U.S. government. Regarding the ICP Program's particular responsibilities, once it has completed its road map with a country, it moves to the status of "sustainment." Obviously, each country will have different conditions and capabilities. However, it is imperative that the ICP Program develop a more robust, multifaceted approach, designing effective techniques to remain in contact with partner countries and monitor the extent to which WMD-detection and response capacity remains intact. Increasing the ICP Program's network with other, similar U.S. government programs is one way to sustain contact with the partner-nation officials. Encouraging partner officials to use the DTRA ICP Program portal is another. The reality is that many newly democratizing states have limited institutional capacity to maintain such indigenous capabilities long term. Cost-effective, long-term means of encouraging these countries to sustain such capabilities are needed.

Use the New International Counterproliferation Program Portal as an Engagement Tool

The ICP Program portal website, introduced in 2010, could, in the future, serve as the centerpiece for sustainment. The portal provides individual online accounts to alumni of ICP Program training activities in an effort to link them into a larger WMD-focused community and to DTRA specifically. The portal can provide a continuous virtual presence in partner nations, even after ICP Program training teams have physically departed the partner country. The portal furthers the ICP Program's mission in a variety of important ways, discussed later in this section.

Dissemination of knowledge is arguably the most important function of the portal, which includes the provision of recent course materials and interactive training modules. Continued education, such as new topics or tactics, techniques, and procedures, as well as refresher courses, are readily accessible and provided on demand. Other potential uses could allow alumni direct access to relevant information databases. All these aspects are especially important for partner countries that have successfully moved beyond intensive program focus and might no longer be receiving frequent hands-on training courses.

The portal should also serve as an important conduit of communication by allowing alumni to easily engage ICP Program officials and provide direct reach back to DTRA experts. This could include the ability to submit and receive answers to counterproliferation questions or to contact key U.S. personnel in the event of a WMD incident. ICP Program staff can also use the portal to monitor some aspects of equipment maintenance and to support partner countries in the replacement of equipment consumables. Through the portal, the ICP Program can inform partner countries of upcoming training courses and events. The portal also has the ability to develop a current list of partner country contacts, mitigating, to a large extent, the issue of whom to call, created by position turnover. In essence, the portal provides course alumni with a vehicle through which to stay regularly informed and engaged in the international community of counterproliferation experts. Given that the portal is a fairly new initiative, it is important that ICP Program managers explore the portal's applications and leverage it effectively as a tool.

Introduce a Recurring Senior-Level Working Group with Partner Countries

Our research found that the ICP Program could benefit from the continued involvement of partner countries' senior officials, beyond the initial engagement at the WMD Executive Seminar. This point arose from discussions with officials from Kosovo in particular. This could be made possible through the establishment of a working group that meets on a regular basis to revisit the objectives laid out at the WMD Executive Seminar. These working-group meetings could be bilateral or, better yet, multilateral, where possible and appropriate, to facilitate regional cooperation and meaningful cross-border discussions. Through recurring meetings, ICP Program action officers and program managers could help to ensure that key officials remain involved in the direction of their countries' involvement with the ICP Program. For example, decisions regarding the appropriate timing for more-sophisticated training could be made with a higher degree of partner-nation input.

Assessment Can Be Strengthened Through Focus on Measurable Objectives and Indicators

Include Objectives, Indicators, and Results in After-Action Reports

The assessment process could be significantly improved by adding rigor to AARs. The RAND study team has examined a sampling of these documents from the case studies, and, although they provide value to future action officers' planning for course logistics, they do not systematically inform the ICP Program managers, or DTRA management, with objective data that would enable the assessment of program performance. Nor do they assist in determining a country's ever-changing requirements, needed to inform the composition and content of future programs.

Currently, ICP Program management practices do not include the systematic compilation of performance outputs, or their assessment, let alone a determination of whether the envisaged country outcomes have been produced. This is an important managerial shortcoming, as the following data should be used to inform improved programs:

- lessons learned from the delivery of a specific course
- whether the curriculum delivered to a country was successfully executed or, just as importantly, if not, why it did not meet expectations
- whether a country's policies or receptivity to ICP Program projects have changed
- the documentation of important official developments in the recipient country.

A reformulation of the AAR development process and its format to provide such data would greatly improve the ability of ICP Program and DTRA senior management to more effectively direct resources and expertise.

Identify Assessment Roles for Stakeholders

DTRA should consider how to establish appropriate roles for assessment among the different stakeholders involved in the ICP Program. Similar to other security cooperation programs, the ICP Program has various stakeholders, each of whom enjoys specific responsibilities and benefits from different points of view with regard to the program's function. These distinct responsibilities, roles, and perspectives naturally incline stakeholders toward different roles with respect to assessment. By decentralizing the assessment process and delegating separate roles, the ICP Program can benefit from each stakeholder's familiarity with specific parts of the program while ensuring a higher level of objectivity with regard to the final outcome. RAND's assessment framework identifies four key roles for stakeholders: data collector, assessor, reviewer, and integrator.

The roles of the many ICP Program stakeholders might look something like those shown in Table 4.1.

The most-relevant data collectors include ICP Program action officers planning the events, ICP Program contractors executing the events, and Country Teams and COCOMs, which liaise closely with partner country contacts in order to obtain relevant information. It is important to note that, although the partner country does not appear in the list in Table 4.1, the partner will be a crucial source of data.

ICP Program managers are the most appropriate body for assessing the data collected, as they have a necessary understanding of the program's goals and the details of how it functions. Therefore, they will be best able to recognize when key data are missing, as well as interpret the data according to their assessment needs. One practical recommendation would be for

Table 4.1
Notional Assessment Roles

Assessment Decision	Data Collector	Assessor	Reviewer	Integrator
Need for the program Design or theory Process or implementation Outcome or impact Cost-effectiveness	Contractor ICP action officer Country Team COCOM	ICP manager	ICP advisory board	OUSD(P) Defense Security Cooperation Agency

DTRA ICP Program managers to convene a series of assessment workshops that include all data collectors to identify the most-relevant indicators and associated output objectives they measure. The assessors could draw from the initial list in this report found in Appendix C, but it is important to keep in mind that indicators should be tailored to help to measure progress relative to specific objectives in a specific country. The assessor—in this case, the ICP Program managers—should be sure to inform data collectors as to precisely what data are needed, relevant to the indicators selected.

A new high-level advisory body, as discussed earlier, would be ideal in the assessment role of reviewer, with both the necessary subject-matter expertise and the required objectivity. Last, OUSD(P), in its role of executive agent of the ICP Program at the strategic policy level, could act as assessment integrator, bringing together insights across the spectrum of security cooperation combating WMD programs managed by DoD.

In addition to assigning explicit roles to aid in the execution of assessments, the ICP Program AARs, as a key source of information regarding program implementation, should be made more detailed and substantive. These documents serve as the vehicle for institutional documentation of program successes and shortcomings. In the future, it will be necessary to change the focus and format of the ICP Program AARs in order to effectively and efficiently implement a more rigorous and robust assessment process.

The Assessment Framework Can Be Applied Beyond the International Counterproliferation Program

Although we tested the assessment framework specifically on the ICP Program for the purposes of this report, the framework could easily be applied to the assessment of other DTRA-managed security cooperation programs. For instance, CTR programs, such as PPP and the Global Initiative to Combat Nuclear Terrorism, as well as the SA/LW, would all be excellent candidates. Moreover, DTRA leadership should consider establishing a working group on program assessments to enhance the effectiveness of such efforts. Such a forum would provide DTRA program managers the opportunity to share best practices and lessons and would facilitate coordination and, ideally, collaboration. ICP Program managers might also consider sharing the results of this ICP Program assessment with the Office of the Deputy Assistant Secretary of Defense for Partnership Strategy and Stability Operations to highlight DTRA's proactive assessment approach to potentially assist OUSD(P) in the development of new, more focused security cooperation assessment guidance in forthcoming versions of the GEF.

Conclusion

The goal of this report is to demonstrate how program-level assessments can be carried out effectively with the help of an assessment framework. The framework itself is general enough to be adapted to a wide variety of programs. Depending on the nature of the program and what aspects of the program most require assessment, there is significant room to tailor the framework to the specific needs of the decisionmaker. Although the framework is designed to aid in the complex process of assessment, it by no means removes all of the inherent challenges involved in assessment, as discussed earlier in this report. Those undertaking assessment will still face obstacles in terms of data availability, the challenge of ensuring objectivity will be ever present, and clear-cut metrics to determine long-term outcomes will likely remain elusive.

However, the application of this framework can aid program managers by providing a construct through which to integrate program objectives and, from these, derive indicators for the five levels of assessment. Once an assessment process becomes institutionalized and stakeholders are identified and assigned specific roles, the process will be made significantly easier. At the very foundation of assessment, the development of measurable objectives is essential. And these measurable objectives must be directly tied to the strategic-level guidance.

This report provides a model for conducting field research according to measurable indicators derived from strategic guidance. It also suggests an approach for identifying the appropriate roles for various stakeholders in the assessment process. Once the assessment is carried out, generating actionable recommendations for improvement will follow more naturally. This report contributes to the established body of work on assessing security cooperation programs and demonstrates the utility of the RAND assessment framework by applying it to the ICP Program. This will, we hope, aid in future efforts to apply the RAND assessment framework to similarly oriented programs, as well as enhance the future effectiveness of the ICP Program and other DTRA- and DoD-managed security cooperation programs.

APPENDIX A

Guidelines for Implementing the Assessment Framework

This appendix lists six concise and specific steps that program managers and their subordinate action officers might consider when implementing the assessment framework discussed in this report. It draws on the recommendations in Chapter Four in a sequential order to facilitate ease of implementation.

1. Emphasize the importance of conducting assessments. During regular staff meetings, introduce the concept of conducting assessments, advising the staff that the ICP Program will use a new assessment process, focused at the country level, with details to follow.
2. Consider holding a one- or two-day workshop for the entire program staff, to introduce assessment concepts and processes. Workshop discussions should provide an overview that addresses, at a minimum, (a) the importance of conducting assessments for the program, (b) how the program fits into broader U.S. government objectives in key countries, (c) the need to connect specific activity objectives to higher-level OSD and COCOM guidance, and (d) how the ICP Program relates to other, similar U.S. government programs (e.g., EXBS, antiterrorism assistance, SLD). A second, more practical discussion might focus on (a) the development of metrics to show outputs and outcomes at the activity level, (b) a discussion of stakeholder assessment roles and responsibilities, and (c) the results of the RAND assessment study in Georgia, Romania, and Kosovo. Finally, a third discussion should focus on process-specific topics, such as assessment roles, responsibilities, timelines, formats, and submission procedures.
3. Assign program staff assessment roles and responsibilities. At the conclusion of the workshop described in step 2, program staff at all levels will have a clear understanding of assessment concepts and processes. In this step, each program staff member is assigned a specific role: data collector, assessor, reviewer, or integrator. Responsibilities for specific assessment actions (i.e., when to collect data, when to submit data) should also be made clear.
4. Institute a regular process of following up on assessment progress. Use regular fora, such as staff meetings, to follow up on assessment progress. Invite staff to present results of progress to date, and invite feedback on improving the assessment process as staff members become more competent in assessments at the country level. Conduct regular debriefing sessions to disseminate assessment results from specific activities to all program staff as a way to apply specific lessons generally across the program.
5. Show how DTRA-managed programs have been combined to support broader DoD and U.S. government objectives at the country level. Hold a division-wide workshop to

consider how the ICP Program is doing with regard to meeting outcome objectives relative to other DTRA security cooperation programs. Ask program managers to present their own assessment results. Identify areas of overlap and security cooperation gaps at the country level.

6. Draft an annual report to capture key insights from the assessment. The report should articulate how the ICP Program is meeting higher-level objectives (e.g., GEF, COCOM). In future iterations, after a more general application of the assessment framework, the report could demonstrate how all DTRA programs contribute to higher-level objectives. In addition, the report could identify best practices and lessons from the year's assessment efforts for potential application to DTRA programs.

APPENDIX B

History of the International Counterproliferation Program

In the 1995 and 1997 National Defense Authorization Acts,[1] Congress directed the Secretary of Defense to develop and implement two counterproliferation initiatives: one between DoD and the FBI and the other between DoD and U.S. Customs. The primary goal was to "expand and improve United States efforts to deter the possible proliferation and acquisition of weapons of mass destruction by organized crime organizations in Eastern Europe, the Baltic countries, and states of the Former Soviet Union."

In 1997, a memorandum of agreement was signed giving DTRA the role of executive agent. Although OSD considered the programs dual track under the ICP Program umbrella, the DoD/FBI program targeted law enforcement officials and their respective agencies, while customs and border officers were the objective of the DoD/Customs program. The initiatives were separately funded and executed until around 2003, when DTRA determined that the efforts would be most effective if administered as a single entity.[2]

Early Years

The ICP Program has undergone numerous changes in its 15-year history. During the early years of the program, the primary goals were as follows:

1. to assist in the continuing establishment of a professional cadre of law enforcement personnel and other officials capable of interdicting and investigating WMD threats and incidents
2. to assist in developing appropriate legislation, laws, regulations, and enforcement mechanisms for deterring, preventing, and investigating WMD threats and incidents
3. to assist in building a solid, long-lasting bureaucratic and political framework in participating nations capable of implementing the above two objectives
4. to establish a long-term and mutually beneficial working relationship between U.S. government agencies and their counterparts in participating states.

Organizationally, the program consisted of three basic elements: policy consultations and assessments, training and technical assistance, and equipment procurement. During the estab-

[1] Public Law 103-337, National Defense Authorization Act for Fiscal Year 1995, October 5, 1994; Public Law 104-201, National Defense Authorization Act for Fiscal Year 1997, September 23, 1996.

[2] Discussion with ICP Program manager, December 2009.

lishment of the program, numerous interagency meetings were held, and, in consultation with the National Security Council, it was decided that initially the program would focus on providing assistance to the community of officials responsible for WMD interdiction in the southern tier of the FSU, particularly Kazakhstan, Uzbekistan, and Kyrgyzstan. The ICP Program then expanded to include the Caucasus and Eastern and central Europe. By fiscal year (FY) 2000, the program moved into the Baltic states. Although some engagement had occurred in such countries as Bulgaria, Romania, Slovenia, and Slovakia, in FY 2001 and FY 2002, the program shifted emphasis to the Balkans, with initial engagement beginning with a conference on defense assistance to counterproliferation and border security in Bled, Slovenia, with representatives from the ten countries of the Southeast European Defense Ministerial.

The key to full engagement with a partner nation in the ICP Program is a DoD requirement to have a government-to-government counterproliferation agreement in place prior to delivery of equipment. OSD has been granted Circular 175 authority by the Department of State to negotiate directly with foreign governments.[3] These WMD agreements not only serve as an umbrella for the ICP Program but are occasionally used by related efforts in DoD, State, and DOE. The agreements represent a government-level commitment to stop the proliferation and trafficking of WMD and related materials, as well as guaranteeing rights, privileges, and exemptions from taxation, among other things, for equipment and personnel. Agreements are in place with Kazakhstan, Uzbekistan, Georgia, Moldova, Romania, Armenia, Azerbaijan, Ukraine, Latvia, Slovenia, Estonia, Lithuania, Croatia, Bulgaria, Albania, Macedonia, and Kosovo. The draft agreement with Kyrgyzstan represents the longest negotiation process, which began in 1997 and has been on and off numerous times with no result.

New Missions

Following the September 11, 2001, terrorist attacks, DoD began to consider substantive changes to the ICP Program. The nexus between terrorism and WMD was added as an integral theme, as cooperation with U.S. partners in the war on terrorism grew in importance for national security. Most scenarios used in tabletop and field exercises focus on terrorist acquisition of WMD, rather than being accident or organized-crime driven, as they had been in the early years. This was especially true with partner countries that did not have direct access to nuclear weapons or materials and did not view WMD as a direct threat to their national security.

Also, two new goals were added to the strategic focus of the program, and renewed emphasis was given to the latter:

- Encourage regional cooperation and information sharing on WMD-related border security and proliferation issues.
- Promote a common understanding of WMD-related threats and risks.
- Establish a long-term, mutually beneficial working relationship between host-country agencies and those of the U.S. government.

---

[3] *Circular 175 procedure* refers to regulations developed by the State Department to ensure the proper exercise of the treaty-making power. See 22 C.F.R. 181.4.

Since 2005, execution of the program has leaned toward practical exercises and response to WMD incidents, moving away from policy and foundational courses. Part of this evolution can be attributed to long-term engagement with some of the partner countries in the FSU, where the program focused in the early years. The second reason is engagement with more-advanced nations that have legislative and bureaucratic foundations to combat WMD but require more-practical applications. In FY 2010, DTRA looked at new courses focused on U.S. national security priorities, including biological weapons and identifying and preventing proliferation pathways.

Expansion of Authority

DoD determined that the geographic expansion into the Balkans did not require additional congressional action. However, when global expansion was considered, OSD submitted a request to Congress for a legislative change. The FY 2005 National Defense Authorization bill allowed global expansion of the program at the direction of the Secretary of Defense after a determination that a "significant (proliferation) threat" exists.[4] Formal expansion of the program awaits OSD approval; however, the ICP Program has supported requests from COCOMs in countries beyond the ICP Program's original geographic boundaries, holding events in Jordan and Singapore.

More recently, in cooperation with NATO's Science for Peace and Security Programme, the ICP Program has expanded its regional efforts initiated in Southeastern Europe by leveraging some of its strongest bilateral ties with long-term partners. Initiatives to engage countries in the Black Sea region have included seminars and a regional exercise. The ICP Program has tried to leverage related efforts, including several Department of State programs and the Proliferation Security Initiative, to create a more cohesive U.S. government engagement with partner countries.

---

[4] Public Law 108-375, National Defense Authorization Act for Fiscal Year 2005, October 28, 2004.

APPENDIX C

Indicators Developed for the International Counterproliferation Program

Using the two documents that lay out the mission and strategy for ICP (OSD Strategic Policy Guidance and Seven-Year Strategy[1]), the study team identified the program objectives (outcome and output) and also the indicators. The documents start by describing the mission for the ICP Program, which we interpret as the outcome objective.

The ICP Program mission (outcome objective) is "to build the capabilities of participating nations to prevent proliferation of weapons of mass destruction (WMD) and related materials across their borders and through their territories." The program's goals (output objectives) are as follows:

1. Enhance participants' ability to prevent, detect, deter, interdict, identify, and investigate the trafficking of WMD-related materials within their territory and across their borders.
2. Promote the establishment of a professional cadre of law and border enforcement personnel dedicated to proliferation prevention, including the participating nation's national security, military, and defense sectors, and encourage effective interagency processes.
3. Emphasize a regional approach to WMD-related border security issues, where feasible.
4. Establish long-term relationships between U.S. government agencies and agencies of participating countries with responsibility for WMD security.

Additional output objectives include the following:

5. administrative requirements
6. individual course objectives
7. long-term funding.

The OSD memo provides goals, which we interpret as output objectives (they are lesser in scope and slightly more measurable). In addition, the team made an assumption that, like with any program, there are administrative and funding output objectives for the ICP Program. In addition, the team assumed that each course has its own output objectives (and that these vary by course). Finally, the team assumed that each of these output objectives contributes to meeting the ICP Program outcome objective. In other words, if the output objectives are successfully met, then one can assume that the outcome objective has been achieved.

The guidance documents also describe deliverables (i.e., tasks)—e.g., plans, funding documents—that can also be thought of as indicators of progress toward objectives. That is, they can be linked to the output objectives. Table C.1 contains these indicators, labeled A

[1] ICP, 2008.

through L, and shows their linkage to the output objectives. The implication for conducting assessments is that the indicators can be analyzed to determine whether the output objectives are being achieved.

In Table C.1, we have added references to the indicators into which we believe the questions give insight.

Table C.1
Indicators and Associated Output Objectives

Indicator	Associated Output Objective
A. Threat assessment	5. Administrative requirements
B. Project plan	5. Administrative requirements
C. Funding document for a new program	5. Administrative requirements
D. Training module and curriculum	5. Administrative requirements
E. WMD agreement	4. Establish long-term relationships between U.S. government agencies and agencies in participating countries with responsibility for WMD security. 5. Administrative requirements
F. Road map for long-term engagement	4. Establish long-term relationships between U.S. government agencies and agencies of participating countries with responsibility for WMD security. 5. Administrative requirements
G. Subject-matter expert assessment of demonstrated proficiency	1. Enhance participants' ability to prevent, detect, deter, interdict, identify, and investigate the trafficking of WMD-related materials within their territory and across their borders. 2. Promote the establishment of a professional cadre of law and border enforcement personnel dedicated to proliferation prevention, including the participating nation's national security, military, and defense sectors, and encourage effective interagency processes. 6. Individual course objectives
H. Knowledge-transfer assessment score	1. Enhance participants' ability to prevent, detect, deter, interdict, identify, and investigate the trafficking of WMD-related materials within their territory and across their borders. 6. Individual course objectives
I. Collection and use of lessons learned	1. Enhance participants' ability to prevent, detect, deter, interdict, identify, and investigate the trafficking of WMD-related materials within their territory and across their borders. 2. Promote the establishment of a professional cadre of law and border enforcement personnel dedicated to proliferation prevention, including the participating nation's national security, military, and defense sectors, and encourage effective interagency processes. 3. Emphasize a regional approach to WMD-related border security issues, where feasible. 4. Establish long-term relationships between U.S. government agencies and agencies of participating countries with responsibility for WMD security.
J. Participation in the ICP Program alumni database	2. Promote the establishment of a professional cadre of law and border enforcement personnel dedicated to proliferation prevention, including the participating nation's national security, military, and defense sectors, and encourage effective interagency processes. 4. Establish long-term relationships between U.S. government agencies and agencies of participating countries with responsibility for WMD security.

Table C.1—Continued

Indicator	Associated Output Objective
K. Funding document for a country's transition into sustainment training	1. Enhance the participant's ability to prevent, detect, deter, interdict, identify, and investigate the trafficking of WMD-related materials within its territory and across its borders. 2. Promote the establishment of a professional cadre of law and border enforcement personnel dedicated to proliferation prevention, including the participating nation's national security, military, and defense sectors, and encourage effective interagency processes. 4. Establish long-term relationships between U.S. government agencies and agencies of participating countries with responsibility for WMD security. 7. Long-term funding
L. Funding document for nontraditional events	1. Enhance the participant's ability to prevent, detect, deter, interdict, identify, and investigate the trafficking of WMD-related materials within its territory and across its borders. 2. Promote the establishment of a professional cadre of law and border enforcement personnel dedicated to proliferation prevention, including the participating nation's national security, military, and defense sectors, and encourage effective interagency processes. 3. Emphasize a regional approach to WMD-related border security issues, where feasible. 7. Long-term funding

In summary, by asking questions, we will get answers that give indications as to how well output objectives are being met, which will, in turn, suggest how well the outcome objective is being achieved. The following describes the mission and goals of the ICP Program and illustrates how elements of the assessment framework are derived from them.

APPENDIX D

Assessment Questions for the International Counterproliferation Program

Table D.1
Assessment Questions

Associated Assessment	ICP Program Partner Questions	Associated Indicator
Need for program		
N1	How were the length and scope of the program determined?	B, F
N2	What was your country's expectation on the program's envisaged length?	B, F
N3	How is progress reviewed, which criteria are used, and do you participate in the review?	G, H, I
N4	How were the courses selected?	B, F
N5	Have you been consulted throughout the execution of the ICP Program as to whether the courses and their associated content are appropriate to your country's conditions and requirements?	B, F, I
Design and theory		
D1	How was your country selected for participation in the ICP Program, and when will continuation in the program be reviewed?	B
D2	Did your country participate in providing threat or risk assessments for the ICP Program personnel?	A
D3	Who in your country conducts the threat or risk assessment?	A
D4	Are threat or risk assessments reviewed and updated, and does your country participate in this review?	A
D5	Is there any disagreement from ministries as to the validity of the WMD threat or risk assessment?	A
D6	Did your country advise the ICP Program managers regarding the ability to absorb training courses?	B, E, F
D7	Have ICP Program staff asked whether their courses are designed and delivered appropriately to your conditions and requirements?	B, F, I
D8	Did your country participate in the design of the ICP Program?	B, E, F
D9	Were you asked to help develop or vet the ICP Program country plan?	B
D10	Are you familiar with the road map for long-term engagement, and, if so, did your country participate in its development?	F

Table D.1—Continued

Associated Assessment	ICP Program Partner Questions	Associated Indicator
D11	Have you been able to discern any differentiation between the road map and the country program plan?	B, F
D12	Has your country been invited to review or comment on the road map?	F
D13	Were your comments and review accepted and implemented by the ICP Program?	
D14	Did your country review the initial project plan?	B
D15	Is your country requested to provide assessments, AARs, feedback, lessons learned, or other tools for improving future programs?	I
D16	Have you seen any corresponding change in programs in light of your input?	I
D17	Have you been invited to assist in the development of WMD-nonproliferation best practices?	I
D18	Has your country been asked to participate in regional conferences regarding WMD challenges, and, if so, have these been effective in assisting your country in reaching the ICP Program objectives?	L
D19	Are you familiar with the ICP Program alumni association, and, if so, do you believe that it would be an effective arrangement?	J
D20	Does your country participate in decisions to move to more-advanced training or equipment?	G
Implementation		
I1	How is progress in the ICP Program determined, and by what method would your country decide not to participate?	G, H, I, K
I2	Is there a method for reviewing the ICP Program's training approach and methodologies when progress is not being made?	G, I
I3	How are the course offerings determined?	B, F
I4	Are course offerings supply driven or demand driven?	F
I5	How, and on what basis, does the ICP Program determine when to repeat course offerings?	G, H
I6	Does your country provide data regarding staff turnover in order to determine when repeat courses should be offered?	F
I7	Does your country participate in the decision to repeat courses?	F, G
Outcome or impact		
O1	Does your country participate in designing short-term and long-term objectives for the ICP Program?	D, F
O2	Does your country participate in assessing increases in WMD capacity that result from the ICP Program activities?	G
O3	Is there an objective assessment of faculty performance following every course?	I

Table D.1—Continued

Associated Assessment	ICP Program Partner Questions	Associated Indicator
O4	Does your country participate in assessing the effectiveness of employed teaching methods?	I
O5	Are courses and support materials updated, and, if so, does your country participate in these updates?	D, I
O6	Are courses designed to meet the requirements and conditions of your country, and does your country participate in their design?	D
O7	Is simultaneous interpretation, or consecutive translation, employed when required, and is it adequate?	D
O8	Are there any the ICP Program metrics to judge progress in WMD awareness and response capacity building as a result of the ICP Program events within your country?	G, H
O9	Is student feedback captured in a systematic way?	I
O10	If so, do you see the results of the student feedback, and does your country assist in using it to improve future training courses?	I

In Table D.2, positive, somewhat positive, and negative are represented by green, yellow, and red, respectively. The spaces where no assessment appears represent either that questions were not asked or that no effective data could be gleaned from the discussion.

Table D.2
Assessments by Country

Assessment	Kosovo	Romania	Georgia
N1			
N5			
D1			
D2			
D3			
D4			
D5			
D6			
D7			
D8			
D10			
D11			
D15			
D18			
D19			
D20			

Table D.2—Continued

Assessment	Kosovo	Romania	Georgia
I2			
I3			
I4			
I5			
I6			
I7			
O1			
O2			
O3			
O4			
O5			
O6			
O7			
O8			
O9			
O10			
C1			

APPENDIX E

Focused Discussions

U.S. Agencies and Organizations

Computer Sciences Corporation
Department of State, EXBS
DHS, ICE
DOE, National Nuclear Security Administration, SLD
DTRA
FBI

Partner Country Agencies

Kosovo

Border Police
Chief Prosecutor
Customs
Department of Emergency Management
Security Force

Romania

Border Police
Ministry of Internal Affairs
National Police

Georgia

Border Police
Customs
Emergency Management
Ministry of Environment, Radiation Safety Department
Ministry of Foreign Affairs
Ministry of Internal Affairs, Counter Terrorism Department
National Security Council

Bibliography

Code of Federal Regulations, Title 22, Foreign relations, Chapter I, Department of State, Subchapter S, International agreements, Part 181, Coordination, reporting and publication of international agreements, Section 181.4, Consultations with the Secretary of State.

Commission on the Prevention of Weapons of Mass Destruction Proliferation and Terrorism, *World at Risk: The Report of the Commission on the Prevention of WMD Proliferation and Terrorism*, New York: Vintage Books, December 2008.

Corso, Molly, "Tbilisi Busts Enriched Uranium Smugglers," Eurasianet.org, April 27, 2010. As of April 22, 2011:
http://www.eurasianet.org/node/60945

DoD—*See* U.S. Department of Defense.

ICP—*See* International Counter-Proliferation Program.

International Counter-Proliferation Program, *Seven-Year Strategy for the International Counterproliferation Program*, Ft. Belvoir, Va.: Defense Threat Reduction Agency, December 2008.

———, working-group meeting on phases, Defense Threat Reduction Agency, September 22, 2009.

Marquis, Jefferson P., Richard E. Darilek, Jasen J. Castillo, Cathryn Quantic Thurston, Anny Wong, Cynthia Huger, Andrea Mejia, Jennifer D. P. Moroney, Brian Nichiporuk, and Brett Steele, *Assessing the Value of U.S. Army International Activities*, Santa Monica, Calif.: RAND Corporation, MG-329-A, 2006. As of April 21, 2011:
http://www.rand.org/pubs/monographs/MG329.html

Moroney, Jennifer D. P., and Joe Hogler, with Benjamin Bahney, Kim Cragin, David R. Howell, Charlotte Lynch, and S. Rebecca Zimmerman, *Building Partner Capacity to Combat Weapons of Mass Destruction*, Santa Monica, Calif.: RAND Corporation, MG-783-DTRA, 2009. As of April 21, 2011:
http://www.rand.org/pubs/monographs/MG783.html

Moroney, Jennifer D. P., Joe Hogler, Jefferson P. Marquis, Christopher Paul, John E. Peters, and Beth Grill, *Developing an Assessment Framework for U.S. Air Force Building Partnerships Programs*, Santa Monica, Calif.: RAND Corporation, MG-868-AF, 2010. As of April 21, 2011:
http://www.rand.org/pubs/monographs/MG868.html

Moroney, Jennifer D. P., Jefferson P. Marquis, Cathryn Quantic Thurston, and Gregory F. Treverton, *A Framework to Assess Programs for Building Partnerships*, Santa Monica, Calif.: RAND Corporation, MG-863-OSD, 2009. As of April 21, 2011:
http://www.rand.org/pubs/monographs/MG863.html

Principal Deputy Assistant to the Secretary of Defense for Global Security Affairs, "Strategic Policy Guidance for the International Counterproliferation Program for Fiscal Years 2010–2015," undated memorandum.

Public Law 103-337, National Defense Authorization Act for Fiscal Year 1995, October 5, 1994.

Public Law 104-201, National Defense Authorization Act for Fiscal Year 1997, September 23, 1996. As of April 25, 2011:
http://www.gpo.gov/fdsys/pkg/PLAW-104publ201/html/PLAW-104publ201.htm

Public Law 108-375, National Defense Authorization Act for Fiscal Year 2005, October 28, 2004. As of April 25, 2011:
http://www.gpo.gov/fdsys/pkg/PLAW-108publ375/html/PLAW-108publ375.htm

Rossi, Peter H., Mark W. Lipsey, and Howard E. Freeman, *Evaluation: A Systematic Approach*, 7th ed., Thousand Oaks, Calif.: Sage, 2004.

U.S. Department of Defense, *National Defense Strategy*, Washington, D.C., June 2008. As of April 22, 2011:
http://permanent.access.gpo.gov/lps103291/2008%20national%20defense%20strategy.pdf

———, "Build the Security Capacity of Partner States," Quadrennial Defense Review Report, Washington, D.C., February 2010, pp. 26–30. As of April 21, 2011:
http://www.defense.gov/qdr/QDR%20as%20of%2026JAN10%200700.pdf

U.S. Government Accountability Office, *Nonproliferation: U.S. Efforts to Combat Nuclear Networks Need Better Data on Proliferation Risks and Program Results—Report to Ranking Member, Committee on Banking, Housing, and Urban Affairs, U.S. Senate*, Washington, D.C., GAO-08-21, October 2007. As of April 21, 2011:
http://purl.access.gpo.gov/GPO/LPS88567

U.S. Joint Chiefs of Staff, *The National Military Strategy of the United States of America 2011: Redefining America's Military Leadership*, Washington, D.C., February 8, 2011. As of April 22, 2011:
http://www.jcs.mil/content/files/2011-02/020811084800_2011_NMS_-_08_FEB_2011.pdf

Weapons of Mass Destruction Commission, *Weapons of Terror: Freeing the World of Nuclear, Biological and Chemical Arms*, Stockholm, Sweden, June 1, 2006.

White House, *National Security Strategy*, Washington, D.C.: White House, May 2010. As of April 22, 2011:
http://permanent.access.gpo.gov/lps90878/national_security_strategy.pdf